GEOMORPHOLOGY
Geomorphic Processes and Surficial Geology

Houghton Mifflin Company, Boston   Atlanta   Dallas   Geneva, Illinois   Hopewell, New Jersey   Palo Alto   London

Robert V. Ruhe
Indiana University

GEOMORPHOLOGY

Geomorphic Processes and Surficial Geology

to Barbara Mills Ruhe
who for many years has helped in the trades
around the world

Cover photo by Andrew Somogy

Printed in U.S.A.

Library of Congress Catalog Card Number: 74–20114
ISBN: 0–395–18553–X

Contents

Preface

The term "process geomorphology" has been around for a long time—in fact, since the beginning of the science. But somehow for many decades a traditional approach in geomorphology developed that was mainly rhetorically descriptive and aimed at the classification of landforms. In recent years interest in geomorphic processes has accelerated, because they not only explain the landscapes on which we live and work, but they in part provide an understanding of the environmental problems that confront us. The objectives of this book are to provide some insight into these processes and to demonstrate their application in attacking both kinds of problems.

Rocks and sediments that lie at or near the surface comprise the internal forms of parts of the landscape. Soils develop on them. All of these features are integral parts of the landscape. They, too, will be approached through processes that may help explain and understand them.

Explanation and understanding are fine from an academic point of view, but someone has to go about developing the explanations and understandings. Methods and techniques must be used, and they will be presented and illustrated by a number of case studies from various places. These examples have been purposely selected from my own work and from the work of close associates during the past several decades. One can more confidently present cases such as these rather than those that are more remote.

Thanks are due to these colleagues, and they will recognize their duly cited studies in this book. Thanks go, too, to those who reviewed the manuscript so carefully: Victor R. Baker, Geology Department, University of Texas at Austin; and W. Thomas Straw, Department of Geology, Western Michigan University. Special thanks are also due to Barbara Ruhe for work on the manuscript.

R. V. Ruhe

GEOMORPHOLOGY
Geomorphic Processes and Surficial Geology

The object of this book is to provide understanding of the nature and evolution of the landscapes and materials at or near the surface of the earth on which we live and work. The first aspect is *geomorphology* (from the Greek *ge*, "earth"; *morphos*, "shape"; and *logos*, "reason") the science of landforms. The second aspect is *surficial geology*, the study of the rocks and mainly unconsolidated materials that lie at or near the land surface. These unconsolidated materials overlying bedrock are *surficial deposits* or *surficial sediments*. They are the materials in which soils form, on which crops grow, on and in which life exists, and in, on, and from which man-made structures are built. Together the landscapes, surface rocks, and surficial deposits comprise a major part of the scenery that is pleasing to the eye and is enjoyed in places set aside for recreation.

Landforms are the features of the earth that together comprise the land surface. They may be large features such as plains, plateaus, or mountains, or smaller features such as hills, valleys, or hillslopes. Most landforms are the products of erosion, but some are formed by deposition of sediments, by volcanic activity, or by movements within the crust of the earth.

Actual construction of a landform can be observed during volcanic activity. An eruption started near Kapoho village on the island of Hawaii in January, 1960, and stopped in February [fig. 1.1(a), (b)], and in that period a volcanic cone and an adjacent lava plain formed. Note the rounding but rugged terrain.

Earth movement may construct sheer and rug-

1

Background and Preparation

ged landforms that extend linearly across country for many miles [fig. 1.1(c)]. When the earth's surface is lowered abruptly, a scarp, or escarpment, is produced. A number of these features in a region make an escarpment landscape.

A *landscape* is a collection or population of landforms. A hill or a valley is a landform; repeated occurrence of hill and valley makes a landscape. A good informal rule is to consider landform as an individual and landscape as a population, keeping both independent of size. Units may be selected and defined as needed.

Certain natural processes form the landscape and surficial sediments, and we will give some attention to the mechanics, dynamics, and chemistry of these processes. Since modern science demands that study be quantitative, a quantitative approach will be used, but purposely only at an elementary level.

The landscape and near-surface features respond to the environment in which they exist. Environment, as well as factors such as vegetation and other forms of life, is dominated by climate. And environment changes through time. All of these relevant factors must be examined, and their impact on land features and processes must be analyzed in order to understand a natural system. A systems-analysis approach helps in deciphering the features and materials of the present and past and aids in estimating possible future changes (fig. 1.2).

Figure 1.1 (a) Kapoho Volcano, Hawaii, during eruption of January–February, 1960. (b) Kapoho cone and adjacent lava plain in 1963. (c) Fault scarp bounding Lake Albert near Mahagi Port, Zaire. (a) and (b) from Ruhe (1969a), by permission. © Iowa State University Press, Ames.

Field in perspective

We will not review the history of geomorphology. That has been done many times (Thornbury, 1954, 1969; Chorley *et al.*, 1964; Fairbridge, 1968). Rather, the intent is to examine briefly the current state of the art and to get on with the job.

Traditional and classic geomorphology is mainly rhetorically descriptive, leading to classification of forms and deductions regarding their origin. Landforms and landscapes are seen as influenced by structure and caused by process,

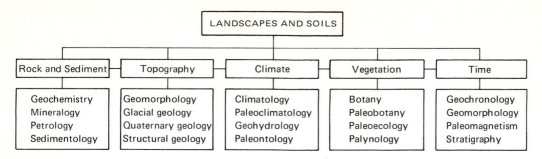

Figure 1.2 Fields of study needed in evaluating factors in a systems analysis of landscapes and soils.

and passing through stages of youth, maturity, and old age. Each stage has a characteristic form appropriate to each geological structure and each geomorphic process. This is the famed Davisian geomorphic cycle (W. M. Davis, 1899a, 1954), which was formulated in the last century and has dominated the field for many decades. The cyclic synthesis, originally applied to erosional landscape, was transposed to glacial, periglacial, karst, and shoreline development (Johnson, 1919). Descriptive analysis was also used for mass movement (Sharpe, 1938). The durability of the classic geomorphology is demonstrated by many textbooks, even some of recent vintage (Cotton, 1941; Thornbury, 1954, 1969; Easterbrook, 1969; Small, 1972).

The traditional approach has been carried over to various kinds of climatic geomorphology even in textbooks, including recent ones (Cotton, 1942; Tricart and Cailleux, 1965; Birot, 1968). This may be an outgrowth of the Davisian "normal" and "humid" development in contrast to "glacial" and "arid" accidents. This separation is maintained in texts, whether in glacial (Flint, 1947, 1957, 1971; Embleton and King, 1968) or in arid geomorphology (L. C. King, 1951).

Following World War II, research in geomorphology broke away from the traditional, and the field of quantitative and dynamic geomorphology exploded. As texts show, achievements have been made in theoretical work (Scheidegger, 1961, 1970), fluvial geomorphology (Leopold et al., 1964; Morisawa, 1968; Carson, 1971), morphometric analysis (Chorley, 1972), and soil geomorphology (Ruhe, 1969a). Although pioneering work on slopes appeared many years ago (Penck, 1924), a text was published only recently (Carson and Kirkby, 1972).

A pertinent analysis concerns the current state of the art of training in geomorphology in 270 colleges and universities in the United States and Canada (White and Malcolm, 1972). Traditional, nonquantitative textbooks are used in about 82 percent of the courses, and correspondingly about 75 percent of the courses are dominantly qualitative, only 25 percent quantitative. The kinds of subjects dealt with in courses reveal in more detail the nature of training (table 1.1). More than 75 percent of the subject matter is from the traditional and classic school, while modern techniques like morphometric analysis and radiometric dating receive scant attention. An objective in the following pages will be to present a more balanced mixture.

Perspective in field

Assuming that one has some knowledge of the many aspects of geomorphology and surficial geology (table 1.1), how does he proceed to study landscapes and surficial deposits? First, since these features exist out-of-doors, he makes extensive study in the field. Having developed concep-

Table 1.1 Subject matter in geomorphology courses in colleges and universities in North America in 1971

Subject matter	Use (%)
Physical and chemical weathering	86
Forms and processes of mass movement	86
Channel patterns and development	85
Flood-plain development	84
Types of terraces	84
Concepts in geomorphology	83
Bed load, suspended load, sediment discharge	80
Water in channels, velocity, discharge, energy	78
Pediments and formation	78
Fluvial geomorphic cycle	77
Glaciers: types, characteristics	77
Dynamic geomorphology	76
Soil formation and soils	76
Mountain and continental glaciation	76
Coast morphology, shoreline erosion and deposition	76
Peneplanation	75
Arid region landscape	75
Karst	75
Stream bed, channel shape	73
Mass movement: hazards and prevention	67
Vegetation, runoff, sediment discharge	66
Soil profiles and description	65
Analysis and retreat of slopes	65
Local geomorphology	65
Mineral chemistry and weathering	64
Hydraulic characteristics of streams	63
Dynamic equilibrium	63
Volcanic landforms	63
Slope erodibility	62
Flood scour of stream bed	61
History of geomorphology	59
Applied geomorphology	57
Criteria of peneplains	54
Terrace correlation	54
Climatic terraces	52
Configuration of sediment on stream bed	50
Patterned ground	47
Flood frequency, rainfall frequency	43
Submarine and ocean-floor topography	39
Mineral chemistry and clay minerals	38
Morphometric analysis	32
Geochronology, radiometric dating	17

From White and Malcolm (1972), by permission of *J. Geol. Educ.*

tual views from prior training and experience, he collects and classifies facts and recognizes their sequence and relative significance. He makes observations and measurements, fitting them into a pattern that reveals their interrelations and predicts the nature of observations not yet made.

Exposures

For a given landscape (fig. 1.3), a topographic map will show the morphology of the terrain, including hilltops, slopes, and drainage lines. The scale of distance and elevation provide a quantitative base for all observations and measurements. If a topographic map is not available or if its scale or contour is not adequate, one must locate all sites of observation and determine their elevations. This can be done crudely by pacing, chaining, and using a hand or Abney level. Accurate measurements are better and can be made by using an engineering transit and a stadia rod or by constructing a topographic map with a plane table and a telescopic alidade survey (R. E. Davis and Foote, 1940).

Along roads and major streams are cuts that permit study beneath the surface. Note that they are sparse and can give only a limited view of the subsurface (fig. 1.3). At the base of each cut, elevations are established at stations along a base line. At stations, contacts between rock, sediments, weathering, soil, faunal zones, or other features in the cut are measured (by hand level, tape, or other optical means) as needed. All measurements are plotted to scale with lateral correlation between stations to produce a quantitative cross section (fig. 2.8). The cross section is then located and oriented as it fits hill crests or slopes in the terrain. Note the road cut just right of the T-road intersection in the sample area (fig. 1.3). It obliquely crosses the hill ridge and hillslopes. The cross section is only a two-dimensional view, but the landscape must be studied in three dimensions.

Figure 1.3 Oblique air photograph of terrain in southern Iowa. Photo by Charles Benn, Iowa State University, Ames, by permission.

Drilling and coring

Drilling is necessary because one must study the entire area. It can be done by hand or by power-driven equipment. Since hand methods require hard manual labor and can be somewhat harmful physically, one should select the proper tool (fig. 1.4). A *soil-coring probe*, pushed into the ground and pulled, gives a reasonably undisturbed core. (Undisturbed cores are the preferred samples.) A

worm auger is screwed into the ground and pulled, which means that the sample is broken from the sides of the hole. *Belgian-type, orchard* (or *bucket*), and *post-hole augers* cut their way into the ground during rotation and are lifted from the hole. If side-hole friction develops, some pulling is required. A little water poured into the hole reduces friction. Depending on the kinds of material penetrated, a depth limit of 20 feet is reasonable for hand drilling. For depths of more than 5 or

Figure 1.4 Hand drilling equipment: (a) soil-coring probe, (b) worm auger, (c) Belgian-type auger, (d) orchard or bucket auger, (e) post-hole auger. Hydraulic drilling equipment: (f) hydraulic soil-coring and drilling machine, (g) hydraulic soil-sampling and foundation drill.

6 feet, a soil-coring probe requires an extension rod. A good combination of drilling tools is an orchard auger with a diameter slightly greater than that of the Belgian-type auger. Where soils are relatively dry, one starts the hole with an orchard auger and takes disturbed samples to a depth where soil moisture increases. Then one shifts to the Belgian-type bit and continues the hole downward. The current cost of the Belgian-type and orchard bits with extension rods and a T-handle to reach a depth of 20 feet is about $100.

Hydraulic drilling equipment is preferable, but it is much more expensive (fig. 1.4). (The present cost of necessary equipment is about $5000.) A hydraulic soil-coring machine capable of extracting 4-foot cores 1 to 2 inches in diameter is an excellent device for shallow work. Where sediments are penetrable, cores can be taken to depths of 40 to 45 feet. One adapts the machine to continuous flight augering by attaching a rotary head and to drive sampling by attaching a cathead.

Deeper work requires a hydraulic soil-sampling and foundation drill, and costs of this equipment and operation mount rapidly. A drill with drilling tools capable of depths of 100 to 150 feet is currently a capital investment of $15,000 to $20,000, and operation of the drill is a two-or three-man job. A good drill is capable of working in rock or unconsolidated sediments by various techniques. In *wash boring*, fluid is forced down the drill rod under pressure and through the rotating drill bit. Drilling fluid with cuttings rises between the drill rod and hole or casing wall to a surface sump, where cuttings are screened and sampled. The depth of the sample relates to the rate of penetration and the rate of rise of fluid in the hole. This disturbed sample is least preferred for studying surficial deposits. Fine earth be-

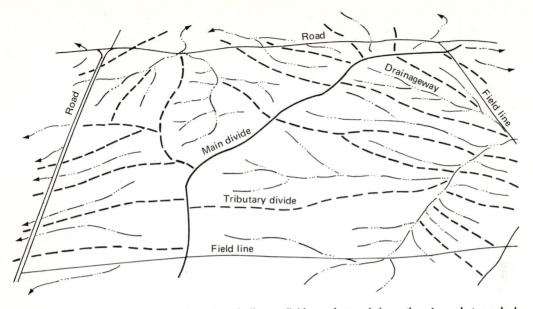

Figure 1.5 Layout of drilling sites along the main and tributary divides on the terrain in southern Iowa photographed in fig. 1.3.

comes mixed as drilling mud, so only clods and rock fragments are useful.

In *flight augering*, continuous spiral sections are added as penetration proceeds downward. The rotating bit cuts earth materials which rise up the flanges of the flights as more or less continuous worms. Sampling is done on the ground at the hole, and the depth of the sample is measured by the rate of return upward. Pouring water in the hole facilitates travel of the sample to the surface. It is important to form a firm, moist worm and not slurry.

Coring is done by three methods. A double-tube swivel-type core barrel is used in *wet rotary drilling*. An outer tube with a cutting bit rotates with the drill rod. An inner tube core barrel on a swivel does not rotate but settles down over the cut core to the depth of tool penetration. Water circulates down the drill rod between the inner and outer tubes of the core barrel through the drill bit and returns to the surface as in wash boring. Different bits, including a diamond one, are used

to core rock or even firm unconsolidated materials.

Hydraulic coring consists of hydraulically forcing a coring tube, such as a Shelby tube, into the ground without rotation to its depth of penetration. The sampling tube is withdrawn hydraulically, and the sample is ejected. Thin-walled sampling tubes give least disturbance to the cores, and the technique is the best for unconsolidated materials but cannot be used in rock. Resistance to penetration can be measured by the hydraulic force required to thrust the sampling tube per foot of penetration.

In *drive sampling* a split-tube sampler is pounded its full length into a material. A standard 140-pound hammer is raised and permitted to fall free 30 inches, pounding the attached drill rod and sampler downward. A blow count is recorded per foot of penetration. The assembly is withdrawn from the hole. The sampler is removed and disassembled, and the core is extracted. For unconsolidated sediments, this

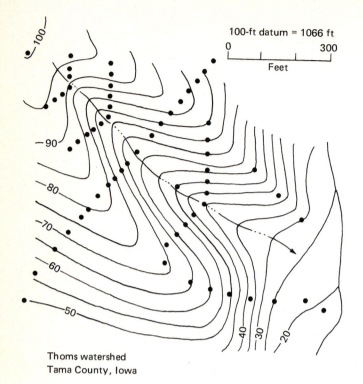

100-ft datum = 1066 ft

0   300

Feet

Thoms watershed
Tama County, Iowa

Figure 1.6 Layout of drilling sites around and across a drainageway. From Vreeken (1968).

technique is preferred next to hydraulic coring.

One can start to recognize subsurface materials and horizons in cuttings and cores by drilling away from a road cut but penetrating all features exposed in the cut. One then correlates by depths and compares the disturbed or core samples with the materials in place in the cut.

To lay out drilling sites in the field, one works from the highland to the lowland or vice versa and attempts to locate and identify all the materials on which the landscape is formed. One penetrates the material beneath the alluvium of the lowland and traces it and all overlying materials under the highland. In the sample area (fig. 1.3), one would drill along the main divide, which is the highest land, and then drill laterally along tributary divides to the lowland and cross it to a main divide on the other side (fig. 1.5). The number of holes required depends directly on the detail desired. Core sections are constructed to scale with ground-elevation control. One correlates from hole to hole with "fence-post" diagrams and lays out the system in three dimensions. The natural model is then available to control and design sampling for other studies.

In many drainageways in the sample area (fig. 1.3), drilling sites would be located on the perimeter of the drainageway descending and crossing the drainage line (fig. 1.6). Note that, in general, drilling traverses should cross contours at right angles. At selected sites, including the highest and lowest locations, drilling should penetrate to a common identifiable layer in the subsurface. This layer is the datum for three-dimensional reconstruction of the system. Where larger landscapes are studied, note the design of drilling traverses to determine the relations within a sequence of stepped land surfaces (fig. 7.17).

Drilling by hand or machine is hard, time consuming, and expensive. In some situations, remote sensing by geophysical means may aid in subsurface study. Small, portable seismographs or electrical resistivity apparatus can be used *if* the geophysical data are correlated with the sequence of features penetrated downward in drilling. The number of drill holes can be reduced, and the stratigraphy between drill holes can be remotely sensed.

Field description

Accurate measurements and descriptions must be made in the field, at a station in a cut or at a drill site in a core. All the results of studies in geomorphology and surficial geology can be no better than these measurements and descriptions.

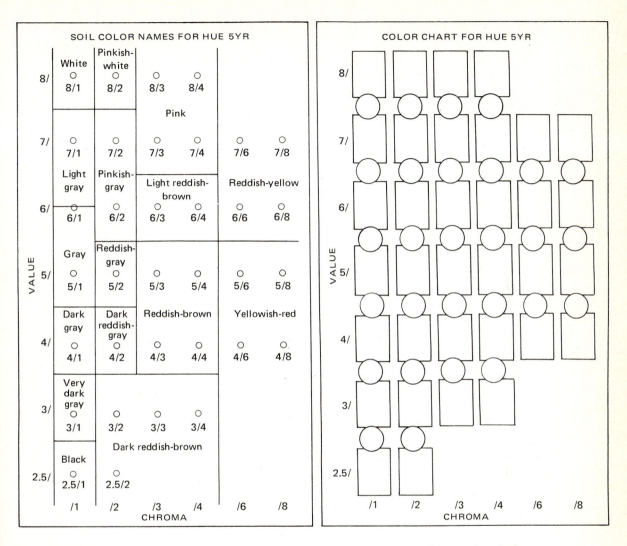

Figure 1.7 Sample sheets of the Munsell Soil Color Chart. The left sheet gives hue, value, and chroma code and color names. The right sheet has corresponding color chips and sight holes. A sample is held behind the sheet at a hole to compare its color with chip, by permission of Munsell Color, Baltimore.

They define the vertical section at a point, and the relation of one section to another defines the three-dimensional natural model. They are also the basis and control for sampling for laboratory study, so the laboratory data produced are no better than the control exercised in field measurement and description. Regardless of the degree of sophistification of data analysis and processing, they too can be no better than the field control.

The depth of layers, horizons, and zones is measured from the ground surface. Thicknesses are also measured, and boundaries are described quantitatively as to distinctness and form, such as smooth, wavy or irregular. In the description one includes color, texture, structure, consistence, reaction, and special features (Soil Survey Staff, 1951).

Color

Colors should be described according to a standard reference like the Munsell Soil Color Chart. Its eight charts display 220 standard color chips

arranged according to hue, value, and chroma. *Hue* indicates relation to red, yellow, green, blue, and purple. *Value* indicates lightness. *Chroma* indicates strength, or departure from a neutral of the same lightness. The symbol for hue is a letter abbreviation of the color, such as R for red, YR for yellow-red, Y for yellow, GY for green-yellow, G for green, BG for blue-green, and B for blue. Each hue is graded from 0 to 10, so 5YR is the middle of the yellow-red hue halfway between 10R (or 0YR) and 10YR (or 0Y). Values range from 0 for absolute black to 10 for absolute white, so a value of 5 is midway between black and white. Chromas begin at 0 for neutral gray and increase in equal intervals from neutral.

The color chart is arranged with hue on a sheet and with value along the Y axis and chroma along the X axis (fig. 1.7). A color notation for hue 5YR, value 5, chroma 6, is 5YR5/6. This color is named yellowish-red—all coded colors in the chart have common names. A sample in the field is visually matched to the nearest color chip in the chart and is named and coded, moist or dry. This standardized system permits comparisons between different workers, which is impossible when nondescript terms such as "buff" or "chocolate brown" are used.

If a material has splotches of color, known as *mottles*, each color is named and coded, and its abundance in the mass pattern is estimated by percent.

Table 1.2 Particle-size classification

Class	mm	μ
Gravel	greater than 2.0	greater than 2000
Very coarse sand	2.0 – 1.0	2000 – 1000
Coarse sand	1.0 – 0.5	1000 – 500
Medium sand	0.5 – 0.25	500 – 250
Fine sand	0.25 – 0.125	250 – 125
Very fine sand	0.125 – 0.062	125 – 62
Very coarse silt	0.062 – 0.031	62 – 31
Coarse silt	0.031 – 0.016	31 – 16
Medium silt	0.016 – 0.008	16 – 8
Fine silt	0.008 – 0.004	8 – 4
Very fine silt	0.004 – 0.002	4 – 2
Clay	less than 0.002	less than 2

From Ruhe (1969a), by permission. © Iowa State University Press, Ames.

Texture

Texture is the particle-size composition of a material, and unfortunately there is a lack of common ground among geologists, engineers, and soil scientists about particle-size categories and texture classes. Sand size, according to geologists, who use the Wentworth scale, is 2 to 0.062 mm, but according to engineers and soil scientists, it is 2 to 0.50 mm. Silt size is 0.062 to 0.004 mm for geologists, 0.05 to 0.005 mm for engineers, and 0.05 to 0.002 mm for soil scientists. Clay size is less than 0.004 mm, less than 0.005 mm, and less than 0.002 mm, respectively. One must adjust for these differences when using data reported by different particle-size scales. A compromise between the Wentworth scale and the scale used by soil scientists is a scale that has five grades of both sand and silt (table 1.2).

Texture-class names are used to describe compositions of sand, silt, and clay particles, but again engineers and soil scientists disagree on the limits and names of classes (fig. 1.8). The texture class *loam* to an engineer has 30 to 50 percent sand, 30 to 50 percent silt, and 0 to 20 percent clay, but to a soil scientist, loam has 23 to 52 percent sand, 28 to 50 percent silt, and 7 to 27 percent clay. These differences must be kept in mind when one uses data from either group. If a material contains gravel-size particles, the adjective "gravelly" or a texture-class name is used, as in "gravelly loam" or "loamy gravel", depending on a component's dominance. Gravel is larger than 2 mm, and it includes pebbles (2 to 64 mm), cobbles (64 to 256 mm), and boulders (over 256 mm).

Structure

Soil structure is the collection of mineral particles into aggregates that are separated from other aggregates by surfaces of weakness. An individual naturally occurring soil aggregate is called a *ped*. Peds are described by shape and arrangement, size, and distinctness and durability, called, respectively, type, class, and grade of structure.

The four types of structure are *platy* (with aggregates generally in a horizontal plane), *prismatic* (with aggregates aligned vertically and bounded by vertical faces), *blocky* (with aggregates organized around a point and bounded on top, bottom, and sizes by flat or rounded surfaces), and *spheroidal*. Columnar structure is prismatic but has rounded tops. Angular blocky

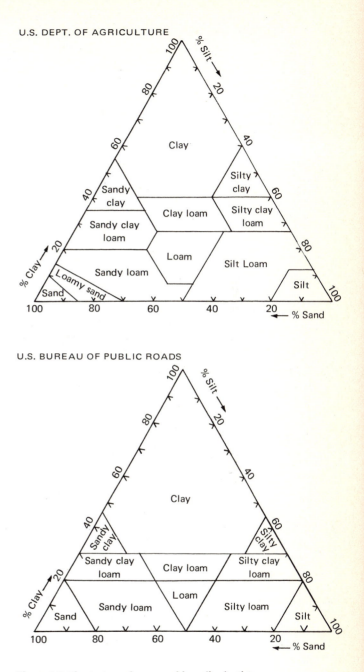

Figure 1.8 The texture classes used by soil scientists and engineers.

Table 1.3 Types and classes of soil structure

Class	Type (mm)			
	Platy	*Prismatic and columnar*	*Angular and subangular blocky*	*Granular*
Very fine	less than 1	less than 10	less than 5	less than 1
Fine	1 – 2	10 – 20	5 – 10	1 – 2
Medium	2 – 5	20 – 50	10 – 20	2 – 5
Coarse	5 – 10	50 – 100	20 – 50	5 – 10
Very coarse	greater than 10	greater than 100	greater than 50	greater than 10

From Soil Survey Staff (1951).

structure has faces intersecting at sharp angles, whereas subangular blocky structure has rounded edges. Spherodial structure is granular.

A material may be essentially structureless or have a weak, moderate, or strong structure. If structureless but coherent it is *massive*; if structureless but noncoherent it is *single-grained*. In weak structure, peds are indistinct, barely visible, and not very durable when one crushes them with the fingers. In strong structure, peds are very distinct and very durable.

Platy structure is measured by one dimension across the plate. Prismatic structure is measured by two dimensions across the vertical axis. Granular and blocky structures are measured by three nearly equal dimensions. These structures are classed by size (table 1.3).

Consistence

Soil consistence is the degree and kind of cohesion or resistance to the shear of a material. When wet or moist, a substance may be sticky or plastic. A material is termed sticky if it adheres to the thumb and fingers when it is rubbed. It is called plastic if it can be formed into a "ribbon" as it is squeezed between thumb and fingers. A substance is friable if it crushes under gentle pressure, and firm if it resists crushing. It is soft or hard depending on the ease or difficulty with which it is broken. Material is indurated if it is in a

cemented state. All of these properties are estimated from grades of weak to strong by hand testing in the field.

Reaction

Reaction is the response of material to a chemical test. One can measure pH with a portable meter or colorimetric kit. Presence of carbonates may be tested with hydrochloric acid; calcareous material will effervesce when drops of acid are applied. Degrees of reaction are recorded from weak to strong. Materials in which free oxides of manganese are present will effervesce with application of hydrogen peroxide.

Other features such as concretions, nodules, veins, and cemented horizons are also described and measured. So that features are not overlooked, one should use a standard field notebook sheet with headings for date, stop or section number, location, elevation, and remarks. For section description, the headings should be horizon, depth, thickness, boundary, and color—moist/dry, texture, structure, consistence, reaction, and special features.

All samples for laboratory studies are indexed to field descriptions at a site and from site to site. This system comprises the three-dimensional framework of the field model.

As the field study progresses, one defines bodies of sediment and components of the land surface by description and measurement. Certain characteristics are grouped to describe a mapping unit whose boundaries are located on the surface and in the subsurface. These limits are recorded on a base map or air photograph as a geomorphic or surficial-deposits map.

Perspective in laboratory and data processing

Modern study requires that laboratory work be coupled with field studies in geomorphology and surficial geology. Kinds of analyses (referred to in many of the following pages) should include physical, chemical, mineralogic, and hydrologic properties as needed. It is beyond the scope of this book to describe the many methods of study, but they are available in numerous manuals and monographs (Black, 1965 a, b; Jackson, 1958; Richards, 1954; Ciaccio, 1973; Soil Survey Laboratory, 1967).

After compiling quantitative data from the laboratory and from the field, one must process, analyze, and present them. The statistical method is the only modern means of handling large masses of numerical data, and the method is *objective* although the results may be affected by *subjective* interpretation. Statistical techniques cannot be presented here, but a few examples are given to help the student understand some of the following pages.

A simple technique is the analysis of the frequency distribution of an array and its central tendency and deviation. A particle-size distribution is a frequency distribution and has a range of sizes (table 1.2). Each size is a part or percentage of the whole. An average sums up or describes the whole; the most important kinds of averages are the arithmetic mean, the median, and the geometric mean. The *arithmetic mean* $\bar{X}$ is calculated by summing all items X and dividing by the number of items N:

$$\bar{X} = \frac{\sum (X)}{N}$$

The *median* is the value of the middle item when items are arranged according to size. For an even number of items, the median is the mean of the two central items. The *geometric mean* is the nth root of the product of n items and is calculated by using logarithms where the logarithm of the geometric mean equals the average of the logarithms of the items. Any average is a single parameter reduced from an array of numerical values and can be used to compare sample against sample. Mean or median particle size is commonly used in sedimentology.

Variation from the arithmetic mean is expressed as *standard deviation, d,* which is the root mean square of differences from the arithmetic mean $X - \bar{X}$, or:

$$d = \sqrt{\frac{\sum (X - \bar{X})^2}{N}}$$

About 67 percent of a frequency distribution falls within one positive and negative standard deviation from the arithmetic mean. In sedimentology the standard deviation is a measure of sorting; the narrower the deviation, the better the sorting.

Table 1.4 Absolute dating of the Tertiary and Quaternary

Era	Period	Epoch	Method	Age
Cenozoic	Quaternary	Recent	C[14]	11,000*
		Pleistocene	K-Ar	1.0 ± 0.5
	Tertiary	Pliocene	K-Ar	1.7 ± 0.4
			K-Ar	12.0 ± 0.5
		Miocene	K-Ar	15.2 ± 0.5
			K-Ar	25.0 ± 1.0
		Oligocene	K-Ar	25.7 ± 0.8
			K-Ar	33.1 ± 1.0
		Eocene	K-Ar	38.0 ± 4.0
			Rb-Sr	55.0 ± 6.0
		Paleocene	K-Ar	57.0 ± 8.0
			K-Ar	59.0 ± 3.0

*Years. The rest of the column is in millions of years.
From Kulp (1961), by permission. © Am. Assoc. Advan. Sci.

Table 1.5 Quaternary stratigraphy of midcontinental United States

Epoch	Stage	Substage		C¹⁴ age (yr)
Recent				
	Wisconsin glacial	Valders*	Valderan [†]	11,500
			Twocreekan	
		Mankato		
		Cary		14,000
			Woodfordian	
		Tazewell		22,000
		Iowan ‡		
		Farmdale	Farmdalian	28,000
			Altonian	(70,000?)
Pleistocene	Sangamon interglacial			
	Illinoian glacial			
	Yarmouth interglacial			
	Kansan glacial			
	Aftonian interglacial			
	Nebraskan glacial			

* After Leighton (1933, 1957). [†] After Frye, Willman, Rubin, and Black (1968). ‡ Defunct by Ruhe et al. (1968).

In data processing, a one-to-one relation may frequently be established between paired measurements. The technique of finding the equation of a curve which passes through or near the points of a set of paired observations is *curve fitting,* and the equation of the fitted curve is an *empirical equation*. The objective is to find the simplest curve that explains the relation, and there are four common types, linear, parabolic, exponential, and power, whose general equations are, respectively:

$$Y = a + bX$$
$$Y = a + bX + cX^2$$
$$Y = ae^{bX}$$
$$Y = kX^n$$

The problem is threefold: (1) determination of the form of the relation (*line of regression*), (2) measurement of the variation about the established form (*standard estimate of error*), and (3) reduction of the measurement of association to a relative base (*coefficient of correlation*). The most widely used method of curve fitting is the *method of least squares*. Standard texts in statistics describe curve-fitting techniques and statistical devices for testing goodness of fit and significance of relations. Numerous applications will be used in the following pages.

More sophisticated multidimensional analysis is complex, and an analyst must resort to a computer. For simpler techniques, small desk-model statistical calculators are now available.

Historical Perspective

Correlation is not only a statistical procedure but also a technique that places landscapes and surficial deposits in proper perspective in space. To understand earth history, one must also consider time. It is generally agreed today that little of the earth's topography is older than Tertiary and that most of it is no older than Pleistocene (Thornbury,

Figure 1.9 (a) Solution pipes filled with red earth in karstic limestone in Bermuda. (b) A thin section of the upper layer of red earth showing light-gray carbonate sand grains in dark clay matrix; an accretionary layer in soil. (c) A thin section of a subsoil horizon in red earth showing absence of primary carbonate grains. (d) A thin section of clastic limestone showing fossil fragments and carbonate grains in calcite matrix; presumed parent of red earth. From Ruhe et al. (1961), by permission of Geol. Soc. Am.

a

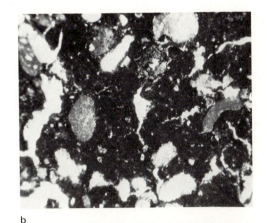

b

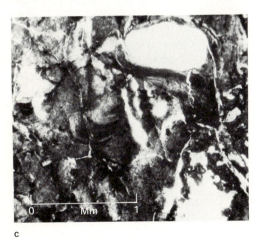

c

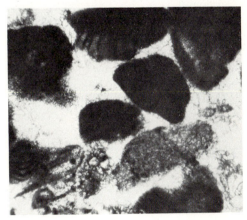

d

Table 1.6 Chemical composition of red earth of Bermuda karst

Horizon*	Depth (in.)	SiO$_2$ (%)	Fe$_2$O$_3$ (%)	Al$_2$O$_3$ (%)	C† (%)	CaCO$_3$ (%)	P$_2$O$_5$ (%)
Ab	0 – 14	11.1	13.2	5.9	0.40	63.1	2.7
B1b	14 – 20	22.8	24.5	12.1	0.25	18.6	9.1
B21b	20 – 26	24.7	31.4	9.2	0.44	3.9	13.4
B22b	26 – 46	20.0	32.6	10.2	0.54	1.9	15.7
B23b	46 – 57	19.1	33.1	9.3	0.43	2.6	16.0
B3b	57 – 64	14.8	28.6	11.1	0.96	14.6	10.8
R	64 +	0.4	0.6	5.2	0.20	92.1	0.8

*See Chapter 2.
† Organic carbon. From Ruhe et al. (1961), by permission of Geol. Soc. Am.

1969); the earth's surface is considered thus to date mainly from the Cenozoic era, which includes the Tertiary and Quaternary periods. Absolute dating (Chapter 11) indicates that little, if any, of the earth's surface is more than 60 million years old, and that most of it is less than 1 to 2 million years old (table 1.4).

The Pleistocene epoch was one of repeated glaciations (Chapter 10), and consequently the history of that time is referred to glacial and interglacial episodes. In the United States, the classic reference section is a composite from the mid-continent region, where four glacial and three interglacial episodes are recognized. The youngest one, the most accessible for study, is consequently subdivided (table 1.5). Radiocarbon dating (Chapter 11) reaches back only part way in this last episode.

In place or out-of-place

Not only are landscapes comparatively young features in the earth's history, but the associated surficial deposits also are relatively young. A continual problem within these young deposits is the distinction between those formed in place by weathering of bedrock (Chapter 2) and those transported and deposited on bedrock by water, wind, ice, and mass movement (Chapters 4-10).

Consider solution pipes in limestone filled with red earth in karst terrain [fig. 1.9 (a)]. To the eye the red earth appears to be *residuum*—the residue from the dissolving bedrock or limestone. But this view is disputed by mineralogic, petrographic, and chemical studies. X-ray diffraction and differential thermal analysis show clay minerals —vermiculite, kaolinite, and amorphous material—throughout the red earth. A calcium aluminum phosphate, crandallite, is also present.

A study of thin sections under a petrographic microscope shows sand-size grains of carbonate in the clay matrix in the upper zones of the red earth but not in the lower zones [fig. 1.9 (b), (c)]. The upper layer must be an accretionary mixture of red earth and carbonate material.

Examine and compare the chemistry of the red earth to that of the underlying limestone (table 1.6). Carbonate content is high in the surface horizon but is very low in the subsoil. The limestone is more than 90 percent carbonate. The upper accretionary layer is shown by the high carbonate content. Note the very low values of silica, iron oxide, and phosphate in the rock, but the large amounts of these compounds in the red earth. Increase from rock to soil is 62X, 55X, and

20X, respectively. The silica ratio means that 31 cubic feet of limestone had to dissolve to leave as residue half a cubic foot of red earth, which forms the *B2lb* horizon. The iron ixode ratio shows that 55 cubic feet of limestone had to dissolve to form one cubic foot of red earth of the *B23b* horizon. For just two of the six soil horizons, a column of limestone at least 86 feet high and one foot square had to dissolve to form the residual products! What thickness of limestone must dissolve to form the other four horizons?

This simple arithmetic points to a problem that arises constantly in explaining "residuum" as a product of weathering in place. Generally there is so much residue that an excessive thickness of rock must decompose to form the product. Other processes must be involved to form the surficial deposits, and these processes, beginning with weathering, will be examined in the following pages.

The weathered rind beneath the land surface, part of any landform, is a result of the physicochemical alteration of rocks or sediments. The group of processes responsible for such alteration is called *weathering*, which requires interaction between mineral matter and atmospheric agents such as wind, temperature, pressure, and water. A solution of limestone provides a simple illustration. Water saturated with atmospheric and soil carbon dioxide reacts with limestone, yielding soluble calcium bicarbonate. Calcite at the solid-liquid interface dissociates into calcium and carbonate ions:

$$CaCO_3 \rightleftharpoons Ca^{++} + CO_3^{=}$$

and reaction between the carbonate ions and dissolved CO_2 forms bicarbonate ions:

$$CO_3^{=} + CO_2 + H_2O \rightleftharpoons 2HCO_3$$

This reaction is the result of three separate equilibria (Weyl, 1958):

$$CO_3^{=} + H^{+} \rightleftharpoons HCO_3^{-}$$
$$H_2CO_3 \rightleftharpoons H^{+} + HCO_3^{-}$$
$$CO_2 + H_2O \rightleftharpoons H_2CO_3$$

So, when water with carbon dioxide reacts with limestone,

$$H_2O + CO_2 + CaCO_3 \rightleftharpoons Ca^{++} + 2HCO_3^{-}$$

calcium and bicarbonate ions are taken into solution. Upon evaporation of the water the reaction reverses, and calcium carbonate is precipitated. In each solution and precipitation cycle, the secondary calcium carbonate is diluted by about one-half with younger carbon dioxide. The mix-

2

Weathering and Soil Formation

ture of younger and older carbon in the secondary carbonate causes a contamination problem in radiocarbon dating of carbonates (Williams and Polach, 1971). Thus, simple weathering complicates radiocarbon dating of soil carbonates.

For the sake of convenience, we may classify weathering processes as physical, chemical, and biological. All of these processes undoubtedly take place contemporaneously, but one of them may be dominant at a site at a particular time.

Physical weathering

Physical weathering is the change of consolidated rock to unconsolidated matter that is caused by physical forces acting on rock in place. In physical weathering stress-strain relations develop. Stress is applied force per unit of area, and strain is mechanical deformation caused by stress. Consolidated rock has inherent strength (either tensile or compressive), which is the resistance to a longitudinal stress that may rupture the material. In unconsolidated sediment inherent resistance to rupture is termed shearing strength. If stress exceeds shearing strength, rocks or sediment breaks down, which is accompanied by a reduction in the size of particles.

Stresses may be generated in rock by the unloading of an overlying mass. When confining pressures are reduced in this way, the release of stresses is mainly vertical. Shear planes form almost at right angles to the direction of the release of stress, and large slabs or sheets break away from the main mass of rock. The fractures generally parallel the ground surface. In granite, as well as in other rocks, curved, concentric slabs may spall from the main mass much like successive layers being peeled from a Bermuda onion. The landform produced thus on rock is like a dome.

The mechanics of this process of *exfoliation* is analogous to that of a coiled spring tightly compressed under a set of weights. As each succeed-

ing weight is removed, the spring rebounds and expands upward along the axis of the coil. Spaces that appear between adjacent whorls are like fractures, and their planar axes are at right angles to the axis of the coil. In addition to unloading, other weathering processes may cause exfoliation.

Weathering agents that generate stresses in materials are the crystal growth of ice or salts, temperature and moisture changes, and organisms. But before they can operate the rock must have initial cracks, fissures, or other openings. Divisional planes such as joints, faults, foliation, bedding, rock cleavage, and other structures provide the necessary access.

A common example of crystal growth is the freezing of water in the cracks of a rock, which can produce stresses greater than the strength of the rock. When water freezes (0°C), there is about a 9 percent increase in volume, and the stress created is almost a ton per square inch. This stress far exceeds the tensile strengths of 75 to 485 pounds per square inch of soils and sediments (Lachenbruch, 1962).

Stresses created by the crystallization of minerals such as gypsum and calcite may also break down rocks. Initial openings or fissures are again required, so that the salt solution may enter the rock or sediment.

Temperature changes due to natural causes are no longer believed to be very effective in rock breakdown. Heating causes expansion of minerals, and cooling causes contraction. Resultant stresses should cause rocks to disintegrate, but laboratory studies by various investigators have failed to verify the importance of these processes in rocks. Contraction upon freezing is, however, responsible for cracking soils and sediments in cold regions (Lachenbruch, 1962).

Wetting and drying cause alternate swelling and shrinking and generate stresses that cause breakdown of material. When a soil clod of given texture, volume, and moisture is dried, the volume of the clod is reduced at the same rate as the volume of water until about 40 percent of the volumetric moisture has been lost (Haines, 1923; R. M. Smith, 1959). Upon further drying (fig. 2.1) the reduction in volume is not as great as the reduction in moisture. Upon wetting, the volume of the clod increases, and at a more rapid rate than it decreases during the previous drying cycle. During another drying cycle, there are similar but not identical volumetric reductions in the clod. Unequal stresses and strains are set up by shrinking and swelling and cause a crumbling of the clod along induced cleavage planes, producing smaller soil aggregates.

Biological agents also break down rocks and sediments. Tree roots growing in cracks pry rock apart, dividing it into small pieces. Ants, termites, and other burrowing creatures tear apart unconsolidated sediments and soils.

All of these physical processes increase the specific surface of a material. *Specific surface* is the sum of the surface areas of all of the particles within a dispersed system of a given volume. Consider a 1-cm cube successively broken into 10 equal volumes (table 2.1). Every time volume is reduced by a tenth there is a tenfold increase in specific surface.

Energy is needed to convert particles to smaller and smaller sizes, and some of that energy is retained in the form of electrical charges on the smaller particles, particularly those of the colloid size (law of conservation of energy). When particles are broken into very small sizes, electrical bonds are broken, which leaves unsatisfied electrical charges on the particles. These particles, whose charges are able to attract or absorb opposite charges, are then ready for chemical processing.

Chemical weathering

Chemical weathering is the change in the chemical composition of rock and sediments that is caused by atmospheric agents, namely water and

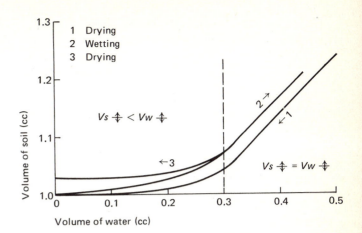

Figure 2.1 The relation of shrinking and swelling of a soil clod to its gain or loss of moisture: Vs, volume of the soil clod including water; Vw, volume of the water. The arrow-bar symbol indicates increase (up) and decrease (down). From Haines (1923).

gases. Four basic kinds of changes take place: additions, removals, transfers, and transformations (Simonson, 1959). The simplest kind of *addition* is hydration, which is the chemical combination of water with a particular mineral. An example of *removal* is the dissolving of calcite from a dispersed system of other mineral grains. Downward percolating water removes the solution product, bicarbonate, from a given place. The solution containing bicarbonate and Ca^{++} ions moves downward, where secondary calcium carbonate may be precipitated, and *transfer* takes place. An example of simple *transformation* is the oxidation of pyrite

$$2FeS_2 + 7H_2O + 15O \rightleftharpoons 2Fe(OH)_3 + 4H_2SO_4$$

with the formation of hydrated ferric oxide and sulfuric acid.

The products of chemical weathering may form a new mineral (the *synthesis mechanism*), or remain as residue when other constituents are removed (the *residue mechanism*) (Jackson and Sherman, 1953). Synthesis involves additions and transformations, but residue results mainly

Table 2.1 Increase of specific surface with reduction in volume

Volume	Specific surface
cm³	cm²/cm³
1.0	6
0.1	60
0.01	600
0.001	6000
0.0001*	60,000

*Colloid size.

from removals and transfers. Water is the main transfer agent in both of these weathering processes, and its activity is fundamentally important.

The role of water in weathering material

When rock breaks down, sediment forms in place, or debris is transported and deposited at a place by agents like water, wind, or ice. These unconsolidated materials are composed of particles of various sizes. When rain or melt water percolates downward into pores or fissures, it reacts in four ways:

1. Chemically combined water, such as the water of hydration, is taken up in minerals and is removed from the hydrology of unconsolidated sediments or soils.

2. Hygroscopic water is absorbed from water vapor by the attractive unsatisfied electrical charges on the surface of mineral particles. Water molecules are dipoles and in the vapor state orient themselves to particles satisfying broken bonds. Hygroscopic moisture is removed from a sample by oven drying at 110°C for 12 hours. The force required to remove hygroscopic moisture is about 30 atmospheres of tension.

3. Capillary water is held by surface tension as continuous films around particles and in capillary pores. This water is removed by air drying or by plant absorption but cannot be removed by

force of gravity. Plants extract moisture and reduce the thickness of the capillary film until the soil capillary force is equaled. When this condition exists throughout the plant-root zone, the plant can no longer extract water, and it wilts. The *wilting point* is at about 15 atmospheres of tension.

4. Gravitational water is water moved by the force of gravity. The amount of water held by a sediment or soil after excess gravitational drainage and after a substantial decrease in the rate of downward movement is termed *field capacity*. The force required to reduce the moisture content at field capacity is 1/3 atmosphere, which marks the lower limit of free or gravity drainage. It can be determined in the laboratory by (1) inserting a sample in a ring, (2) saturating the sample, (3) weighing the sample, (4) placing the sample on a porous plate and subjecting it to 1/3 atmosphere in a pressure cooker for several days, (5) reweighing, (6) oven drying, and (7) reweighing. Excluding step (3), field capacity can be measured. Including step (3), gravitational water (or free water) can be measured gravimetrically or volumetrically.

The volume of water forced from the sample under 1/3 atmosphere of pressure is about the same as the volume of the gravity-drained pores. The volume of water remaining after 1/3 atmosphere of pressure but removed by oven drying represents the volume of the capillary pores. Therefore

$$\frac{\text{volume of water at less than 1/3 atmosphere}}{\text{volume of saturated ring sample}}$$
$$\times\ 100 = \text{aeration porosity \%, and}$$
$$\frac{\text{volume of water at greater than 1/3 atmosphere}}{\text{volume of saturated ring sample}}$$
$$\times\ 100 = \text{capillary porosity \%}$$

Aeration and capillary porosities depend upon the particle-size distribution and the packing or arrangement of particles within a material (fig. 2.2). In order to understand particle-void relations, visualize equal-sized spheres packed in a container, one above another and side by side. The volume of solids is 52 percent, and the volume of voids is 48 percent. Now, if the spheres are arranged in a pyramid, the volume of solids occupies greater space—74 percent of the total volume. This change is the effect of different packing. If a container is packed as originally, but with smaller equal-sized spheres in the original void space, the volume of solids increases, and pore space decreases. This is the effect of varying the particle size.

If voids are relatively large and connected, free water will readily flow through the sediment. If the voids are fine and connected, capillary water movement will dominate. The ability of a medium to transmit a fluid is its *permeability*, and it involves both kinds of water movement. The relation of texture, porosity, and permeability is self-evident for two contrasting media (table 2.2).

The movement of water is extremely important in chemical weathering. In the case of limestone, solution depends on the rate at which water enters the medium and becomes saturated with Ca^{++} and $CO_3^=$ ions. This rate is in turn governed by the rate of transport of solute away from the solid-liquid interface, where the solution is always saturated (Weyl, 1958). After the solute is saturated, no further solution can take place unless the solvent is moved by both free and capillary flow. Unsaturated solvent must be brought to the liquid-solid interface to keep the weathering process active.

Refresher in elementary chemistry

In the foregoing discussion, electrochemically charged atoms and radicals, or cations and anions, were mentioned. Cations carry positive charges (Ca^{++}); anions carry negative charges ($CO_3^=$). The combining capacity of an ion is its valence, which is expressed as a multiple of the combining capacity of the standard hydrogen ion (H^+). Common ions are released and can react

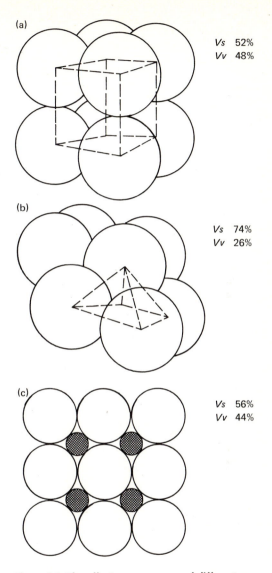

(a)   Vs   52%   Vv   48%

(b)   Vs   74%   Vv   26%

(c)   Vs   56%   Vv   44%

Figure 2.2 The effect on pore space of different arrangements of particles: Vs, volume of solids; Vv, volume of voids.

Table 2.2 Relation of texture, porosity, and permeability of contrasting media

Particle size				Porosity			Permeability
Sand (%)	Silt (%)	Clay (%)	Aeration (%)	Capillary (%)	Total (%)		(in./hr.)
10.0	62.7	27.3	19.1	37.4	56.5		24.0
2.6	45.0	52.4	5.6	45.8	51.4		1.7

From Ulrich (1950).

in chemical weathering (table 2.3). H^+ and OH^- combine to form water. H^+ combines with other anions to form acids (for example, H^+ and Cl^- form HCl, hydrochloric acid). Cations combine with OH^- ions to form hydroxides or bases (Na^+ and OH^- form NaOH, sodium hydroxide). Cations combine with anions, excluding H^+ and OH^-, to form salts (Na^+ and Cl^- form NaCl, sodium chloride).

During chemical weathering, simple and complex minerals decompose, yielding ions and simpler compounds. Solution of calcite is an example of simple dissociation. In more complex feldspar minerals the release of compounds is orderly. During electrodialysis, following fine grinding of microcline ($KAlSi_3O_8$) and albite ($NaAlSi_3O_8$) in water, compounds are removed in the order $Na_2O > K_2O > SiO_2 > Al_2O_3$ (Armstrong, 1940). Reaction of acid-washed, silt-sized feldspar particles with acidic clays causes loss of cations in the order anorthite > bytownite > labradorite > andesine > oligoclase > albite (Graham 1941, 1950). This sequence is the familiar calcic, calcic-sodic, sodic-calcic, sodic-plagioclase series. The general loss of cations is Ca > Mg > Na > K.

Particles of minerals, particularly clay minerals, also carry electrical charges. The clay crystal carries a negative electrical charge, which permits it to hold positively charged cations, such as H^+, Ca^{++}, Mg^{++}, K^+, and Na^+. If the clay holds many H^+ ions the clay is acid. If the clay holds many ions capable of forming bases such as Ca^{++}, Mg^{++}, K^+, Na^+, and NH_4^+, the clay is alkaline.

The cations held by clay minerals are exchangeable: H^+ may displace Na^+; Ca^{++} may displace H^+; and so forth. *Ion exchange* is the reversible process by which cations and anions are exchanged between solid and liquid phases or between solid phases. Cations are more important than anions in exchange processes; hence, the terms *cation exchange* or *base exchange* are commonly used. *Cation exchange capacity* (CEC) is the ability of a medium to hold exchangeable ions. The exchange capacity is measured and expressed in milliequivalents (me) per 100 grams of sample. One milliequivalent is one milligram of hydrogen or its equivalent, and a milligram is 0.001 gram. In general, the amounts of exchangeable ions in a material vary with the kind of clay mineral, the amount of clay, and the amount and kind of organic matter.

A medium is acid, neutral, or alkaline depending upon its concentration of H^+ or OH^- ions. At neutrality the H^+ ions equal the OH^- ions. For example, a liter of distilled water is neutral and contains 1×10^{-7} gram of active hydrogen; its normality is $1 \times 10^{-7}N$. This small fraction is awkward, however, so the normality is expressed as the logarithm of the reciprocal of the hydrogen-ion concentration. A reciprocal of a number is the inverted fraction, so the reciprocal of 1×10^{-7} is 1×10^7, whose logarithm is 7. The logarithm of the reciprocal of the hydrogen-ion concentration is expressed by pH. Hence, a pH of 7.0 is neutral on a scale ranging from 0 to 14. As the pH decreases in steps of 0.5 from 7.0 to 4.5, a medium is termed very slightly, slightly, moderately, strongly, or very strongly acid respectively. Similarly, a medium is described as very slightly to very strongly alkaline as the pH increases from 7.0 to 9.5. During chemical weathering, acidity is caused by removal of bases and accumulation of hydrogen. Alkalinity is caused by accumulation of bases.

During chemical reaction the pH of the medium controls, in part, the solubilities of specific minerals (Marshall, 1964). Calcite in water with dissolved CO_2 at pH 7.0 has a relatively low solubility of about 0.01 g/l. However, in an acid medium at pH 3.9, the solubility is greater, about 1.1 g/l. In contrast, amorphous silica has low solubility of 0.2 g/l in an acid medium at pH 2.0. In an alkaline medium at pH 10.6, solubility is higher, 1.1 g/l.

As noted previously, bases are released from silicate minerals generally in the order Ca > Mg > Na > K. During electrochemical bonding, ions are held in weathered products generally in the order H > Ca > Mg > K > Na. The exchange complex tends to be dominated by H and Ca, whereas Na tends to be removed during the weathering process. The relation of the exchangeable bases and hydrogen is the *base saturation*, which is expressed as

$$\frac{Ca + Mg + Na + K}{Ca + Mg + Na + K + H} \times 100$$

$$= \text{base saturation \%}$$

Chemical weathering indices

If common rock-forming minerals such as orthoclase—the potash feldspar ($KAlSi_3O_8$)—or anorthite—the calcic plagioclase ($CaAl_2Si_2O_8$)—weather chemically, the products should be potash (K_2O), lime (CaO), silica (SiO_2), and alumina (Al_2O_3). These compounds can be measured in the laboratory, and various ratios of them may be arranged as indices of the intensity of chemical weathering (Jackson and Sherman, 1953). Some of the common ratios are:

$$\frac{SiO_2}{Al_2O_3} = \text{silica : alumina ratio}$$

$$\frac{SiO_2}{Fe_2O_3} = \text{silica : ferric oxide ratio}$$

$$\frac{SiO_2}{Al_2O_3 + Fe_2O_3} = \text{silica : sesquioxide ratio}$$

$$\frac{K_2O + Na_2O}{Al_2O_3} = \text{alkali : alumina ratio}$$

Table 2.3 Common ions in chemical weathering

Cations			Anions		
(+)	(++)	(+++)	(−)	(=)	(≡)
H	Ca	Al	OH	CO_3	PO_4
K	Cu	Fe	Cl	HPO_4	
Na	Fe		HCO_3	SO_4	
NH_4	Mg		H_2PO_4		
	Mn		NO_3		
	Zn				

$$\frac{CaO + MgO}{Al_2O_3} = \text{alkaline earth : alumina ratio}$$

$$\frac{CaO}{MgO} = \text{calcic : magnesia ratio}$$

$$\frac{K_2O}{Na_2O} = \text{potassic : sodic ratio}$$

Alumina is often used as the basis for calculating the relative weathering losses of other compounds. The alkaline earths are removed more readily than alumina, and so comparison may be made between the ratio of alkaline earth to alumina in the original material and the same ratio in the weathered product. The relation of calcium to magnesium is commonly expressed as a simple Ca:Mg ratio.

Other ratios are also used to evaluate the degree of weathering. A molar ratio of CaO/ZrO_2 (Beavers et al., 1963) shows the calcium-bearing mineral hornblende is more susceptible to weathering than the more resistant zirconium-bearing mineral zircon. Lower ratios indicate relatively more intense weathering than higher ratios. Compare this molar ratio with a mineral-weathering index hornblende/zircon.

Biological weathering

The impact of organisms on weathering is profound. Plants grow on the land, and their roots

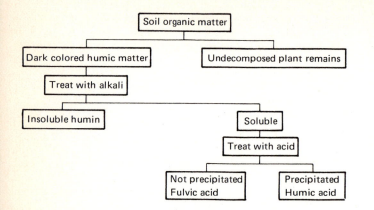

Figure 2.3 **The components of soil humus defined on the basis of simple chemical extraction.**

penetrate the soil. Animals live on and in the soil. At the close of a life cycle, plant and animal residues are returned to the soil and are decomposed by microorganisms such as bacteria, actinomycetes, fungi, algae, and animal protozoa (Stevenson, 1964).

Microorganisms

Bacteria are the most abundant microorganisms in soils and may vary in number from a few to many millions per gram. The population depends on temperature, moisture, aeration, pH, and the organic and inorganic food supply. Bacteria are autotrophic or heterotrophic. Autotrophs get their carbon from carbon dioxide and their energy supply from the oxidation of inorganic substances. They oxidize N, S, Fe, Mn, H, and CO to assimilable forms. Heterotrophs oxidize organic compounds for their energy supply and decompose cellulose, hemicellulose, starch, sugar, protein, fat, wax, and other substances. Some of the autotrophs are "nitrifying" organisms, which oxidize ammonia to nitrites and then nitrites to nitrates, which are then available to plants as nutrients. Some of the heterotrophs are "denitrifying" organisms, which compete with plants for available nitrates. Some of these nitrates are converted to "fixed" nitrogen during biological denitrification and are unavailable for plant use.

Actinomycetes, which are heterotrophs, are next in abundance to bacteria, and they increase in number during later stages of decomposition of organic matter. Their main role may be to break down very resistant plant and animal materials. They also liberate and recirculate nitrogen and other nutrients.

Fungi are heterotrophic, and they also decompose organic matter. They are unicellular yeasts, mushrooms, and free-living molds; they have high acid tolerance and capably decompose cellulose, hemicellulose, pectin, starch and lignins. They also transform nitrogen and synthesize many complex compounds.

Algae are less abundant than the other forms. Since they are normally the invaders in barren soils, their residue is often the base for colonization of other forms.

Protozoa are predators, and they feed on other microorganisms. Their role may be to maintain a natural balance in the system.

The main biological role of the organisms discussed above is to transform organic matter. The organic system contains many compounds, including organic residues that are decomposing, and it also contains metabolic by-products of organisms. Carbohydrates and related compounds, proteins and their derivatives, lignins, fats, and tannins are among the various decomposition products. The end product is *humus*, which is a complex mixture of amorphous and colloidal substances formed from modified plant materials and synthesized microbial tissues (Stevenson, 1964).

Organochemical nature

The chemistry of soil organic matter is one of the most complex systems in nature. However, generalization is useful in understanding it. If humus is treated with an alkali, some of the material is soluble, and some is insoluble (fig. 2.3). The insoluble fraction is *humin* (Mortensen and Himes, 1964). The soluble fraction, treated with acid, precipitates *humic acid. Fulvic acid* remains in solution but may be extracted with further treatment. The fractions are mixtures of many kinds of organic chemical compounds; none is a distinct chemical entity. This classic fractionation procedure is a common pretreatment of samples for radiocarbon dating and will be discussed in Chapter 11.

These organic complexes are extremely active chemically. Cation exchange capacities of humic acids range from 170 to 590 me/100 g at a pH range of 4.5 to 8.1 (Rydlerskaya and Tisckenko, 1944). In comparison, an active clay mineral such as montmorillonite may have a cation exchange capacity of 60 to 150 me/100 g. The ability of humic acid to complex metals is exceedingly great. As the acid moves readily in weathering processes, transfer of organometallic complexes may be expected.

Mineral weathering

Physical, chemical, and biological processes break down rocks and minerals to smaller sizes and alter them so that some weathering products may combine to form new minerals or may remain as residue when other products are removed.

Stability of minerals and mineral weathering indices

The stability of a mineral species during weathering depends on its hardness, cleavage, flaws in the crystal, solubility, and particle size or specific surface. Stability is determined by the mineral's persistence within a material through time, its geographic correlation with factors of weathering intensity (usually climate), its persistence within a particle-size or specific-surface distribution, and its persistence in depth within a zone of weathering (Jackson and Sherman, 1953). In addition, under controlled conditions in the laboratory, various mineral species show little effect—while others show great effect—from a given chemical treatment.

The weathering sequence of common rock-forming silicate minerals with the least stable minerals at the top is shown in table 2.4 (Goldich, 1938).

In a given environment olivine and calcic plagioclase weather more readily than succeeding members of the sequence. Quartz is the most resistant to weathering. The general order of release of bases can be compared with this mineral weathering sequence.

Some of these minerals are "light minerals"

Table 2.4 Weathering sequence of common rock-forming minerals.

Olivine (Mg, Fe)	Calcic plagioclase (Ca, Al)
Augite (Ca, Mg, Fe, Al)	Calcic-alkali plagioclase (Ca, Na, Al)
Hornblende (Ca, Na, Mg, Fe, Al)	Alkali-calcic plagioclase (Na, Ca, Al)
Biotite (K, Mg, Fe, Al)	Alkali plagioclase (Na, Al)
Potash feldspar (K, Al)	
Muscovite (K, Al)	
Quartz (Si)	

From Goldich (1938).

such as quartz and feldspars, while others are "heavy minerals." (These terms relate their specific gravities.) They may be distinguished by whether they float or sink in a liquid whose specific gravity is 2.9. Within each of the two separates, mineral species are identified and the number per sample is counted. A light mineral weathering ratio (Wrl) is the percentage of the sample within a specific particle-size fraction:

$$Wrl = \frac{quartz}{feldspars}$$

This ratio is useful in comparing the *relative* mineral weathering of similar sediments in local areas on land surfaces of different ages (Ruhe, 1956b). This ratio does not indicate "absolute" weathering (Brewer, 1964).

Numerous stability series have been given for heavy minerals (Pettijohn, 1941; Dryden and Dryden, 1946; R. Weyl, 1952; Jackson and Sherman, 1953). There is general agreement that olivine, amphiboles (hornblende), and pyroxenes (augite) are the least stable and that zircon and tourmaline are the most stable in weathering. A heavy mineral weathering ratio (Wrh) is:

$$Wrh = \frac{zircon \ and \ tourmaline}{amphiboles \ and \ pyroxenes}$$

This ratio is also useful in comparing the relative mineral weathering of similar sediments in local areas on land surfaces of different ages (Ruhe, 1956b). It has been used to determine the effect of variations in the texture of sediment on the intensity of weathering (Brophy, 1959). Where $Wrh =$ (zircon + tourmaline)/hornblende or (zircon + tourmaline)/garnet, the ratios showed relatively greater weathering in coarse-grained, open-textured glacial outwash than in glacial till. In a weathering zone in outwash about 90 percent of the hornblende and 70 percent of the garnet has been removed, but in glacial till only 60 percent of the hornblende is gone, and garnet is little changed.

In working with mineral stability ratios, more than one particle-size fraction must be examined. Mineral-species content varies with particle size. Note the differences in heavy mineral content in two particle-size grades in a soil and in a lower parent material (table 2.5). To compare soil and loess, depletion of hornblende is shown by the ratio Z + T/H, regardless of particle size. By

Figure 2.4 Diagrammatic cross sections of layer-silicate clay minerals. Compiled from Jackson (1964).

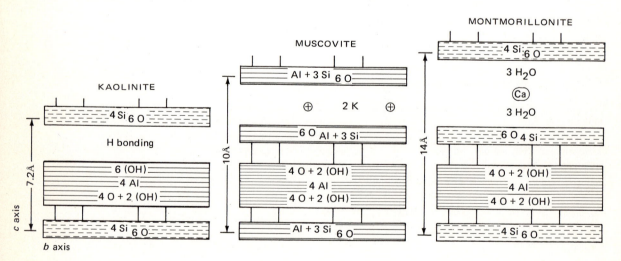

particle size, however, magnitudes differ from 13:1 to 7:6.

Clay minerals

Clay minerals are layer silicates that are very important in weathering. Their particle size is generally less than $2\,\mu$ (less than 0.002 mm). They are mainly responsible for the physical properties of plasticity and structure and the chemical property of cation exchange. Because of their large specific surface, the rate of weathering increases in species that are more stable as larger particles. Minerals other than layer silicates also occur as clay-size particles and are important parts of the clay fraction of sediment or soils.

The layer-silicate minerals may be visualized as sandwiches (fig. 2.4). One layer consists of the silicon tetrahedron in which a Si^{4+} ion lies equidistant from four oxygen ions, or SiO_4^{4-}. Another layer consists of an Al, Fe, or Mg ion bonded to six equally spaced oxygen or hydroxl ions, which form an octahedron as $Al(OH)_6^{3-}$. Tetrahedral silica sheets attached to octahedral hydroxyl sheets form the layered structure, and various combinations of these two sheets form the different layer-silicate minerals. Where there are alternating pairs of tetrahedral and octahedral sheets, as in kaolinite, the mineral is a 1:1 layer silicate. Where the octahedral sheet is bounded by two tetrahedral sheets, as in montmorillonite, the mineral is a 2:1 layer silicate. The 1:1 and 2:1 structural units may be stacked like pages in a book. Note the small size, measured in angstrom units ($\overset{\circ}{A}$ = one ten-millionth of a millimeter).

Adjacent structural units may be bonded by cations, such as hydrogen in kaolinite and potassium in muscovite (fig. 2.4). Adjacent units may also be weakly bonded so that water and exchangeable cations may be interlayered, as in montmorillonite. The first type has a nonexpandable lattice, but the second type expands, which is typical of swelling clays.

Table 2.5 Heavy mineral content in different particle-size grades in Sangamon soil and Loveland loess in southwestern Iowa

Mineral	Sangamon soil		Loveland loess	
	100-50μ (%)	50-20μ (%)	100-50μ (%)	50-20μ (%)
Epidote	24	29	18	25
Tourmaline (T)	2	3	4	3
Zircon (Z)	4	11	1	11
Garnet	10	10	5	9
Hornblende (H)	45	20	54	23
Titanium minerals	10	20	10	18
Apatite	3	5	4	9
Others	2	2	4	2
Z + T/H	0.13	0.70	0.09	0.61

From Ruhe and Cady (1967).

Other layer silicates such as chlorite, halloysite, illite, and vermiculite are common in soils and sediments. Illite is a disordered form of muscovite. Mineral species are usually interstratified, and a mixed-layer lattice results.

Because of the small size of these minerals, they must be studied with x-ray or electron microscope techniques. X-ray diffraction patterns are most commonly used.

Clay-size minerals and clay minerals also differ in weathering stability. In a relative index of weathering (table 2.6), higher numbers represent greater resistance to weathering. In a group of clay-size minerals in a soil or weathering profile, the frequency percentage of each mineral is plotted against the weathering index scale of 1 to 13. As intensity of weathering increases, frequency distributions are displaced toward higher index numbers (Jackson, 1964).

Clay minerals are very active components in weathering. Their small size is responsible for a very large specific surface within a medium. They are negatively charged and have a wide range of exchange capacities (me/100 g): kaolinite, 3–15; illite, 10–40; chlorite, 10–40; montmorillonite,

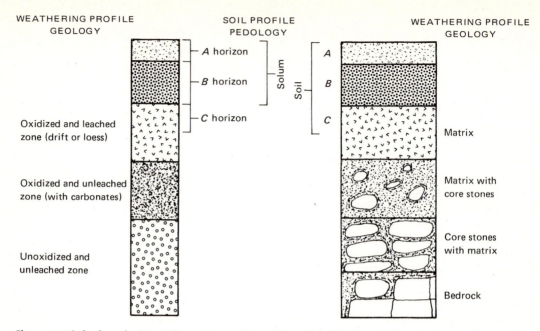

Figure 2.5 Kinds of weathering profiles and comparisons of the soil profile. From Ruhe (1969a), by permission. © Iowa State University Press, Ames. Also from Ruxton and Berry (1957), by permission of Geol. Soc. Am.

60–150; and vermiculite, 100–150. Cations of the clay minerals are exchanged readily with organic colloids and metallic ions as an important part of the weathering process.

Weathering zones and weathering profiles

As physical, chemical, biological, and mineral weathering progresses, alteration of rocks and sediments develops three-dimensionally beneath the land surface. Layers or zones of altered material are formed parallel or subparallel to the land surface. The intensity of weathering decreases in each succeeding lower layer. A *weathering zone* is an altered layer that differs physically, chemically, and mineralogically from the layers adjacent to it and that extends laterally beneath the land surface. A *weathering profile*, a vertical section from the land surface downward, crosses the

weathering zones to the unaltered material. Usually a soil and a soil profile are just beneath the land surface in the uppermost weathering zone.

Vertical section

The weathering profile is studied to determine the nature and arrangement of weathering zones at a site. The first site is selected beneath a level or nearly level hill summit, so that erosion of the uppermost zone is almost precluded. Colors, including mottled patterns, are described according to a reference such as the Munsell Soil Color Chart. Texture, structure, and consistence of material also are described. If special features, such as nodules, concretions, sheets, tubules, and the like, are present, they are described, and their compositions are identified. Organic matter is located and described. Rocks and minerals are identified.

Simple chemical tests are then performed. The pH is determined colorimetrically or with a portable pH meter. Free carbonates are estimated by using HCl and noting slight, moderate, strong, or violent effervescence.

All of these properties are combined in and associated with a specific zone in the vertical section. Combinations of properties differ in the zones above and below. The boundaries of the different layers can therefore be set and described; thickness of each layer can be measured. The weathering profile has then been outlined.

Weathering zones are defined in various ways. In the upper Mississippi Valley region there is a common weathering profile in glacial, eolian, and other sediments (fig. 2.5). A soil profile is beneath the surface in the uppermost zone, which is the *oxidized and leached zone*. This zone is yellowish-brown, and it has a rusty appearance (hence the term "oxidized"). There are no carbonates detectable by acid test, but because the next zone does have carbonates, the upper zone is "leached." The lower zone, which has similar color and carbonates, is *oxidized and unleached.* The next zone is dark gray, but "oxidation" (the yellowish-brown color) extends downward in it along fractures from the overlying zone. Because of its gray color, the lowermost zone is considered *unoxidized,* and because it contains carbonates, it is *unleached* (Ruhe, 1969a). The vertical arrangement is the weathering profile. The soil solum at the top is designated zones I and II, and the zones just described are III, IV, and V (Brophy, 1959). (See fig. 2.6.)

Where bedrock is weathered, another system may be used (Ruxton and Berry, 1957; Berry and Ruxton, 1959). Zones are established according to the percentage of core stones and the percentage of weathered matrix (fig. 2.5). Core stones, derived from the bedrock, dominate in the zone above partially weathered bedrock, and they are surrounded by weathered debris. In the next higher zone, the weathered debris occupies a greater volume than that of the core stones, which

Table 2.6 Stability in weathering of common clay-size and clay minerals (13 = greatest stability)

Index Number	Mineral
1	Gypsum, halite
2	Calcite, dolomite, aragonite, apatite
3	Olivine, hornblende, pyroxenes, diopside
4·	Biotite, glauconite, chlorite
5	Albite, microcline
6	Quartz
7	Illite, sericite, muscovite
8	Vermiculite, 2:1 mixed lattice-clay minerals
9	Montmorillonite, 2:1 mixed lattice-clay minerals
10	Kaolinite, halloysite
11	Gibbsite, boehmite, allophane
12	Hematite, goethite
13	Anatase, zircon, rutile, ilmenite, corundum

From Jackson (1964) in *Chemistry of the Soil*, ed. F. E. Bear, by permission. © Van Nostrand Reinhold Co., New York.

decrease in number nearer the top of the zone. The next zone is almost wholly weathered debris with few, if any, core stones. A soil profile caps the vertical section. The zones are numbered downward, and the vertical section is the weathering profile.

When accurately described and measured, the weathering profile is the foundation for sampling for laboratory studies. It is a systematic outline for the location of samples in relation to each other. Physical, chemical, and mineralogical properties analyzed in the laboratory are meaningful within this organized system and can be interpreted properly. In a weathering profile in glacial till in Illinois, clay less than 2μ accumulates in zones I, II, and the upper part of III, and carbonates are leached from all three zones (fig. 2.6). Heavy mineral weathering ratios show depletion of hornblende in zones I and II. Clay minerals are altered at much greater depth and in the calcareous material. Chlorite changes to vermiculite-chlorite in zones III and IV and completely to vermiculite in zones II and III. Illite becomes mixed layer

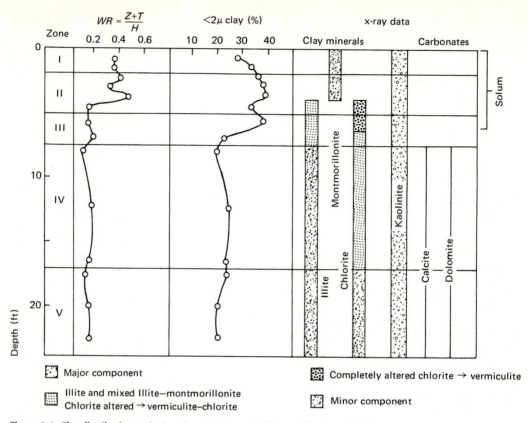

Figure 2.6 Clay distribution and mineral zoning in weathering profile in glacial till in Illinois. From Brophy (1959), by permission of Ill. Geol. Surv.

illite-montmorillonite in zones II and III. Alteration to montmorillonite occurs in zone I and the upper part of zone II. From these data it is clear that the most intensive weathering is in zones I and II and the upper part of zone III. The soil profile occupies this vertical space (compare fig. 2.5). Beneath the soil, weathering decreases as depth increases.

X-ray diffraction patterns of samples from a weathering profile show progressive change in clay mineralogy from the lowest zone upward through the higher zones and into the surficial soil (fig. 2.7). Note the upward decrease in illite and the change in 2:1 lattice clay. Referring to the weathering stability sequence of clay minerals (table 2.6), one may devise a weathering ratio

from x-ray diffraction patterns. Kaolinite and iron-rich chlorite are identified by the 7.2Å peak, and weatherable illite is marked by the 10Å peak. The diffraction intensity (DI) ratio is (Willman, Glass, and Frye, 1966):

$$\frac{\text{x-ray diffraction intensity in (counts/sec) 10Å spacing}}{\text{x-ray diffraction intensity in (counts/sec) 7.2Å spacing}} = \text{DI ratio}$$

The alteration of chlorite is shown by a decrease in intensity of the 7.2Å spacing and by an increase in the DI ratio. The value of the DI ratio increases upward in the profile until the alteration of illite (shown by a decrease in intensity of the 10Å spacing) begins. The value of the DI ratio de-

creases upward in the profile as illite is more severely altered. Reversal in the DI ratio indicates
the relative amount of depletion of illite and chlorite and also the intensity of weathering.

Lateral variation

Studying variations of weathering zones aids in
interpreting the formation of landforms. Cuts or
drill cores, when properly located, permit study
in three dimensions. Weathering zones in a railroad cut provide a cross section through a ridge in
loess in southwestern Iowa (fig. 2.8). Note that
boundaries between weathering zones approximately parrellel a lower buried soil and subparallel the surface in the center of the cut. Also note
that the zones outcrop on the left and right hillslopes. Since the zones pass through the cut, they
must have formed in the subsurface; they could
not have formed on the hillslopes with one edge
exposed to the atmosphere. Their outcrop is the
result of beveling by hillslope erosion, and the
hillslopes are younger than the weathering zones.
The zones, used stratigraphically, aid in interpreting the origin of the landform.

Weathering zones may be traced cross-country
for many miles (fig. 2.9). Note that two deoxidized zones, when traced from site 50 to site 4,
merge as one thicker zone. Hillslopes that bevel
the zones at all sites may be fitted into a hillslope
erosion episode.

Weathering zones can be used to study not
only weathering per se but also geomorphic and
stratigraphic relations. Another pertinent use is
the fit of soils to the zones (fig. 2.8). Note that
specific soils of the soil envelope fit specific
zones and thicknesses of zones.

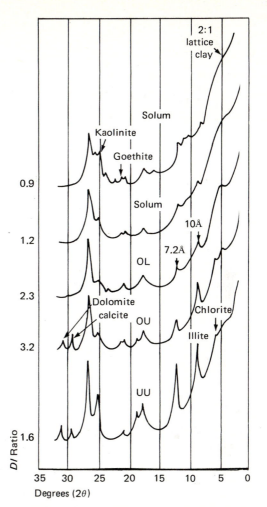

**Figure 2.7 X-ray diffraction patterns of samples from
a weathering profile at Funkhouser section in Illinois: OL, the oxidized and leached zone; OU, the oxidized and unleached zone; UU, the unoxidized and
unleached zone; DI, the diffraction-intensity ratio.
From Willman et al. (1966), by permission of Ill.
Geol. Surv.**

Soils

Soil has many definitions. On the basis of properties, soil is the outer layer of the earth's crust, usually unconsolidated, ranging in thickness from a
mere film to more than 10 feet; it differs from the

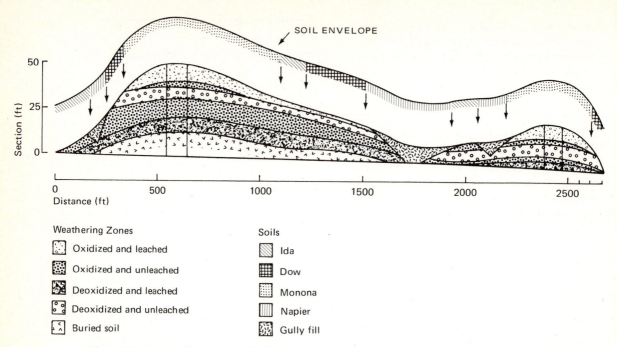

Figure 2.8 Lateral variation of weathering zones beneath a ridge in southwestern Iowa. The soil envelope has exploded. From Ruhe (1969a), by permission. © Iowa State University Press, Ames.

material beneath it in color, texture, structure, physical constitution, chemical composition, biological characteristics, chemical processes, and morphology (Marbut, 1951). In terms of its relation to plants, soil is the natural medium for plant growth on the surface of the earth and is composed of organic and mineral materials (Yearbook of Agriculture, 1938). Genetically, soil is the collection of natural bodies on the earth's surface "that support plants and that have properties due to the effects of climate and living matter, acting upon parent material, as conditioned by relief, over periods of time" (Soil Survey Staff, 1951). As a system, soil is the portion of the earth's crust whose properties vary with the soil-forming factors:

$$s = \int (cl, o, r, p, t, \ldots)$$

where s is soil; cl, climate; o, organisms; r, topography; p, parent material; and t, time (Jenny, 1941).

All of the preceding definitions leave something to be desired. How are soils separated from weathering zones? Any separation must be arbitrary; however, one possibility is to make two weathering categories: (1) formation of parent material from rock, and (2) formation of soil from parent material (Robinson, 1949).

The term *soil* is a generalization. *A soil* is something specific. It is a three-dimensional body that occupies part of a landform. In working with a soil, a parcel must be defined and delineated on the ground. Vertically, a soil must extend from the land surface into the parent material. Laterally, a soil must extend far enough so that its field properties may be measured and sampling may be done (Cline, 1949). Laterally, a soil will change to another unit with dissimilar properties. The boundaries of each soil may be mapped.

How is a soil delineated out-of-doors? Cores are extracted by hand or by power-driven devices at many places in a parcel of ground. Properties are measured and described for each core and site and are arranged according to layers, whose thicknesses are measured. Two soils are established, A and B (fig. 2.10). In coring from A to B, differences become evident. Up to a certain boundary on the ground, profiles are more similar to A; beyond the boundary, profiles are more similar to B. From a parcel composed mainly of A, the soils grade to a parcel composed mainly of B. A transitional zone, marked by an arbitrary boundary, welds the two end members A and B on the landscape. Correlation of many profiles from site to site in the field is necessary if a three-dimensional body is to be delineated.

Soil profile

The *soil profile*, a vertical section of a soil at a site, is a representative sample of a soil and may be relatively simple or complex (fig. 2.11). It is composed of *soil horizons*, which are layers approximately paralleling the land surface and whose properties have been determined by soil-forming processes. Biological weathering is important in producing organic matter on and in the uppermost horizons, and the depth of rooting of plants

Figure 2.9 Lateral variation of weathering zones at successive topographic-divide sites in southwestern Iowa. From Ruhe (1969a), by permission. © Iowa State University Press, Ames.

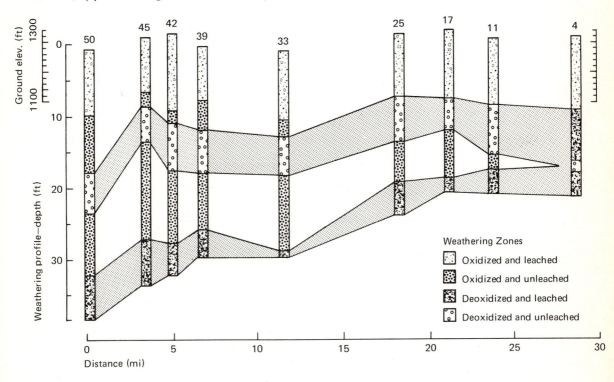

Weathering Zones
- Oxidized and leached
- Oxidized and unleached
- Deoxidized and leached
- Deoxidized and unleached

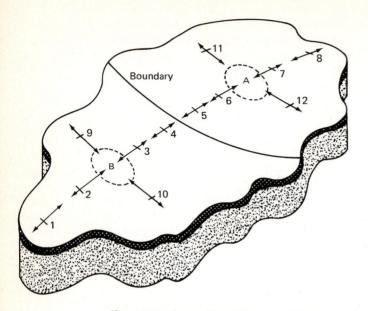

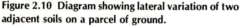

Figure 2.10 Diagram showing lateral variation of two adjacent soils on a parcel of ground.

tends to control the biological activity of the soil-forming processes.

The horizons of the soil profile are defined by association of the properties specified in Marbut's definition of soil. Master horizons are broadly divided into organic and mineral horizons. They are symbolized alphabetically with upper-case letters and Arabic numerals, *O1, A1, B2,* and so forth. Each master horizon may be further subdivided into *A11, A12, A13,* or *B21, B22, B23,* and so forth, if greater detail is needed.

The *A* and *B* horizons are the *solum*, which is the part of the soil formed by soil processes. The base of the solum, or the lower boundary of the *B3* horizon, separates the soil from the remainder of the weathering profile (fig. 2.5).

Soils have different kinds of profiles. If soils form in the same kind of parent material, with similar topographic positions, under similar climates, and during the same time but under different vegetation, their profiles and properties differ

distinctly (fig. 2.12). A Tama soil formed under grass has a thick, dark *A1* horizon that grades downward into a lighter-colored *B* horizon with soil structure and accumulated clay. The *B* horizon grades downward into the *C* horizon. The horizon sequence downward is *A1, A3, B1, B2, B3,* and *C*.

A Fayette soil formed under trees has a thin, dark *A1* horizon over a light gray, ashen, platy *A2* horizon, which grades downward into a darker-colored *B* horizon with soil structure and accumulated clay. The *B* horizon grades downward into the *C* horizon. The horizon sequence downward is *A1, A2, B1, B2, B3,* and *C*.

Note the distinct differences in properties of soils from similar positions in the soil profiles (table 2.7). Removal of constituents from the *A* horizon and transfer to the *B* horizon has been more intense under forest than under grass. This is the eluvial-illuvial process of soil formation (fig. 2.11). Note that the ratios (*B/A* horizon) of clay and iron are 2.27 and 1.8 under forest and 1.04 and 1.13 under grass. The forest soil is more acid (lower pH) and bases have been depleted with lower base saturation. Note the difference in intensity of weathering indicated by CEC. The impact of different vegetation results in distinctly different soils (fig. 2.12).

There are many different kinds of soil profiles, and a few more kinds will be examined to illustrate the effects of other soil-forming factors. In arid regions carbonates are so important that carbonate layers have been recognized as a master soil horizon, the *K* horizon (Gile et al., 1965). In New Mexico, a Red Desert soil is formed in rhyolite gravel [fig. 2.13(a)]. Note that clay (less than 2μ), organic carbon, and Fe_2O_3 increase from the *A2* to the *B2* and into the upper part of the *K* horizon and then decrease in the lower parts of the *K* horizon (table 2.8). Note the high content of carbonate in the *K* horizon, which is cemented and indurated. This material is known geologically as calcrete or caliche.

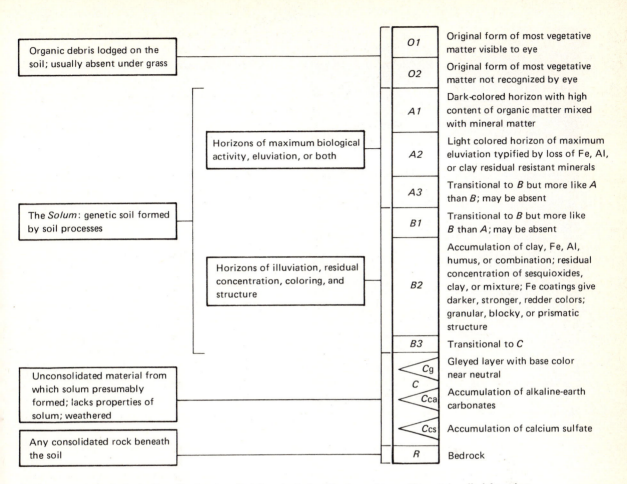

Figure 2.11 A hypothetical soil profile showing all of the principal soil horizons. No profile contains all of them, but every profile has some of them. Modified from Soil Survey Staff (1951).

The carbonate is responsible for unusual soil properties. The infiltration rate of the *K21* horizon is 0.05 inch per hour, and its air dry unconfined compressive strength is 7880 pounds per square inch (Gile, 1961). This horizon is essentially impermeable and exceptionally strong. Consider what effects this horizon has in a landscape in regard to penetration of rain water, inducement of runoff, and resistance to erosion. Soil properties can have a considerable impact on other geomorphic processes.

This profile is also an excellent example of the polygenesis of soil. Clay, iron, and organic carbon could not have been formed in or transferred downward into the almost impermeable and indurated *K* horizon. These compounds must have been in the layer prior to cementation of the *K* horizon by calcium carbonate. Thus, two episodes of soil formation are represented: (1) the clay-iron-humus profile formed earlier, and (2) the carbonate profile formed later.

In tropical areas an entirely different kind of

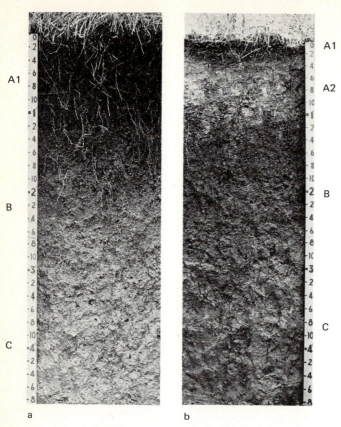

Figure 2.12 **Tama soil formed under grass (a). Fayette soil formed under trees (b). Scales in feet. Soil horizons symbolized.** Photograph by R.W. Simonson, by permission.

kaolinite, hematite, and goethite in the Latosol indicates a high intensity of weathering.

Indurated iron oxide horizons in a soil, usually called laterite [fig. 2.13 (c)], also disrupt penetration of rain water, induce runoff, and resist erosion. A common term for such horizons in soils is *pan*, which may be clay pan, a horizon with high clay content with great density and impermeability; iron pan, a horizon of laterite or ferricrete; silica pan, a horizon of silcrete; and so forth. All of them can cap hill summits and form relatively resistant ledges on the shoulders of hillslopes.

Soil classification

Because there are different soil profiles and soils, they may be classified. The criteria used refer directly to soil horizons and include number, color, texture, structure, relative arrangement, chemical composition, and thickness. Additional criteria, not restricted to soil horizons, are thickness of the true soil (the solum), character of the (parent) soil material, and geology of the soil material. To these 10 criteria of Marbut (1922), mineralogy of soil horizons and of parent material should be added.

According to the arrangement of these properties, categories are assigned from lowest to highest rank: (1) A *soil series* is a collection of soils that have genetic horizons with similar properties and arrangement in the soil profile. A soil series usually is formed from a particular parent material and occupies an area on the landscape (Riecken and Smith, 1949). (2) A *soil family* is composed of various soil series, grouped together for some purpose, such as to emphasize the properties important for the growth of plants. The family category thus has use-of-soil bias. (3) *A great soil group,* or great group, is a collection of soils, soil series, and soil families whose profiles have a particular set and arrangement of soil horizons. This category is used to assess the impact of environment on soils, especially that of vegetation and climate. The effect may be broad, geographic

soil profile forms. It contains large amounts of iron oxide, dominantly hematite, which causes a red color. A soil from Hawaii (table 2.9) called a Latosol [fig. 2.13 (b)] is essentially clay (less than $2\,\mu$), almost exclusively kaolinite, and sesquioxides which also are colloidal size. Note the thorough weathering expressed by low base saturation, low pH, and low CEC in the B and C horizons, in spite of the fact that this soil formed from basalt or detritus derived from basalt. The soil has a high iron oxide content of hematite and goethite. According to the weathering sequence of clay-size minerals (table 2.6), an assemblage of (such as a humid or subhumid climatic belt), or it may be local (such as a well-drained hillcrest next to poorly drained bog).

The soil series and the great soil group are probably the most useful classification categories. The difference between them may be illustrated by the Tama and Fayette soils in Iowa (fig. 2.12, table 2.7). A collection of soils whose properties are similar to the Tama soil and whose horizons are *A1, A3, B1, B2, B3,* and *C* would be classified as the Tama soil series. Soils similar to the Fayette soil would form the Fayette soil series. At the great soil group level, note the difference in sets and arrangement of soil horizons between the Tama and Fayette soils. Where Tama has a

A1, A3, B, and *C* set and arrangement, Fayette has *A1, A2, B,* and *C* set and arrangement. Thus, these two soil series are placed in different great soil groups. The Tama is a *Prairie Soil,* and the Fayette is a *Gray-Brown Podzolic Soil.* These terms are names previously used for certain great soil groups.

The great groups are organized into higher taxonomic units: suborder and order. Recently, a new soil classification system has been introduced in the United States (Soil Survey Staff, 1960). All soils have been placed in 10 orders (table 2.10). The new system is extremely technical, has very complex terminology, and should be left to the professional soil taxonomist. Many

Table 2.7 Comparison of properties of Tama and Fayette soils of Iowa

Horizon	Depth (in.)	Clay less than 2μ (%)	pH (1:1)	Base saturation (%)	Organic carbon (%)	CEC (me / 100g)	Fe$_2$O$_3$ (%)
Tama soil under grass							
A11	0 – 6	28.6	5.7	66	2.35	20.6	0.9
A12	6 – 11	32.2	5.8	62	1.95	21.4	1.5
A3	11 – 16	34.2	5.7	68	1.42	23.8	1.6
B1	16 – 20	35.6	5.8	72	0.97	24.0	1.6
B21	20 – 25	35.4	5.7	73	0.68	23.6	1.7
B22	25 – 29	33.2	5.7	74	0.45	22.8	1.7
B23	29 – 35	30.5	5.7	76	0.34	21.7	1.7
B31	35 – 45	28.2	5.8	77	0.21	20.9	1.7
B32	45 – 51	28.5	6.1	81	0.15	20.3	1.6
C	51 – 61	27.6	6.5	84	0.12	20.1	1.5
Fayette soil under forest							
A1	0 – 2	13.0	5.6	56	5.63	21.1	0.9
A2	2 – 9	13.5	5.0	20	0.66	7.3	1.0
B1	9 – 17	16.9	5.0	34	0.30	8.4	1.2
B21	17 – 24	23.8	5.0	47	0.20	13.0	1.6
B22	24 – 35	27.7	4.9	51	0.16	17.1	1.7
B23	35 – 42	30.7	5.0	64	0.17	19.9	1.8
B3	42 – 48	29.6	5.2	65	0.14	19.1	1.8
C	48 +	26.3	5.1	66	0.12	17.5	1.8

From Ruhe (1969a), by permission. © Iowa State University Press, Ames.

Figure 2.13   Red Desert soil in rhyolite gravel in New Mexico (a). The prominent carbonate horizon is light-colored. Latosol composed essentially of kaolinite and sesquioxides formed from basalt debris in Hawaii (b). Note the slip planes. Indurated laterite in Hawaii (c). Scales in feet. From Ruhe (1967), by permission of N. Mex. Bur. Mines and Miner. Resour.

other systems of soil classification exist, and most of them have a national origin, such as the system for Australian soils (Stace et al., 1968). An international system is available through the United Nations (Dudal, 1968).

Soil landscapes

If soils are part of the landscape, the pattern of soils on the ground is the *soil landscape*. It may be large and occupy a young glacial drift surface of almost 12,500 square miles, as one in Iowa does. This soil landscape is called a *soil association area* (fig. 2.14). Three dominant soils, Clarion, Nicollet, and Webster, are on hill summits, hillslopes, and lower-lying closed depressions in the poorly drained landscape. The surface soil horizon darkens in color and thickens in the order of Clarion, Nicollet, Webster, and Glencoe. The

subsoil becomes grayer and less permeable and its texture becomes heavier. When these soils are aligned in a traverse with their changing properties, they form a *toposequence*. The sequential graying and change in permeability of the subsoil demonstrate a sequential change in hydrology. Interaction takes place between topographic control and internal drainage of the soil.

Where the effect of one soil-forming factor causes changes in others, a soil landscape system evolves which is the soil catena (Milne, 1936a). A *soil catena* is a physiographic complex of soils or a sequence of soils between the crests of hills and the floors of adjacent depressions or drainageways whose profiles change from point to point in the traverse depending on conditions of drainage and past history of the land surface.

Two catenary variants are recognized. In one, the topography is modeled by erosion and other processes, from a material originally similar in

Relationship of slope, vegetation, and parent material to soils of the Clarion-Nicollet-Webster soil association area

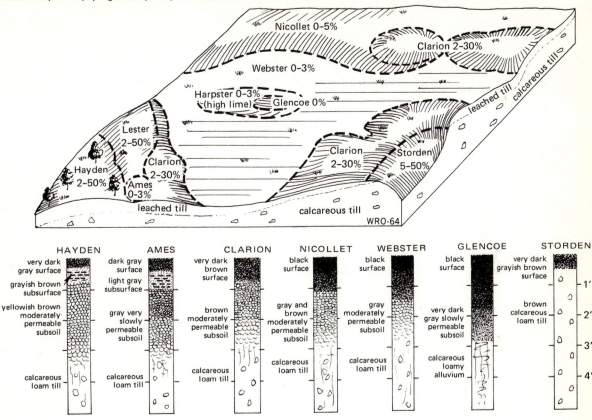

Figure 2.14 A soil landscape on the Cary glacial drift in Iowa showing the relation of soils, vegetation, and parent material to soils of the Clarion-Nicollet-Webster soil association area. Note the representative profiles of the soils. From Oschwald et al. (1965).

composition at all levels at which it is exposed. Soil differences are developed by drainage conditions, differential transport and deposition of eroded material, and leaching, translocation, and redeposition of mobile chemical constituents. In the other, the topography is carved from two or more materials, of which the uppermost is just beneath the hilltop, while the lower is exposed successively downslope. A geological factor is added to the other conditions and causes soil differences.

Milne's catena is an ideal model for working with soil landscapes on hillslopes. It effectively integrates geomorphic processes of erosion, transportation, deposition, hydrologic effects due to different drainage conditions, and pedological processes. It also considers the past history of the land surface. These are facts fundamental to soil geomorphology, and they are extremely useful in applied studies out-of-doors.

In the final analysis, the weathering of rock and the formation of soil prepares the landscape for erosion and transport and deposition of eroded debris.

Table 2.8   Properties of Red Desert soil in southern New Mexico

| Horizon | Depth (in.) | Whole soil greater than 2 mm (%) | Less than 2 mm fraction | | | | |
			less than 2 μ (%)	Organic carbon (%)	Fe₂O₃ (%)	CEC (me / 100g)	Carbonate (%)
A2	0 – 2	39	13.6	0.29	1.1	12.6	tr.
B21	2 – 6	53	17.8	0.37	1.2	13.6	tr.
B22ca	6 – 11	60	22.4	0.82	1.2	17.1	10
K21	11 – 12	45	23.5	0.73	1.3	16.3	80
K22	12 – 25	74	20.0	0.34	1.2	16.4	58
K3	25 – 50	65	13.3	0.14	1.0	10.1	23

From Gile, Hawley, and Grossman (1970).

Table 2.9   Properties of the Paaloa soil of Hawaii

Horizon	Depth (in.)	pH (1:1)	Organic carbon (%)	Fe₂O₃ (%)	CEC (me / 100g)	Base saturation (%)
A1	0 – 11	5.7	3.39	29.6	28.5	23
B21	11 – 19	4.9	1.62	31.4	28.8	1
B22	19 – 30	4.7	1.24	28.6	23.9	1
B23	30 – 41	4.8	0.76	20.0	17.6	2
B3	41 – 60	5.0	0.49	23.0	19.2	4
C	60 – 66	5.1	0.37	19.3	17.1	4

Table 2.10 Soil orders, approximate great soil groups, and weathering intensity

New order	Mnemonic	Approximate old great soil group	Dominant clay minerals
1. Entisols	Recent	Azonal soils, Humic-Gley soils	Montmorillonite, mica
2. Vertisols	Invert	Grumusols	Montmorillonite
3. Inceptisols	Inception	Ando, Sol Brun Acide, Brown Forest, Humic-Gley soils	Allophane, mica, inter-stratified layer silicates
4. Aridisols	Arid	Desert, Reddish Desert, Sierozem, Solonetz, Solonchak, Brown, and Reddish-Brown soils	Mica, vermiculite, inter-stratified layer silicates, chlorite
5. Mollisols	Mollify	Chestnut, Chernozem, Prairie, Rendzina, Brown, Brown Forest, Solonetz, Humic-Gley soils	Montmorillonite, mica, vermiculite, chlorite
6. Spodosols	Podzol	Podzols, Brown Podzolic soils, Ground-Water Podzols	Sesquioxides, inter-stratified layer silicates, 2:1/2:2 intergrades, mica
7. Alfisols	Pedalfer	Gray-Brown Podzolic, Gray Wooded, Noncalcic Brown soils, Chernozem, Planosols, Half-Bog soils	Mica, montmorillonite, 2:1/2:2 intergrades, chlorite, kaolinite
8. Ultisols	Ultimate	Red-Yellow Podzolic, Reddish-Brown Lateritic, Half-Bog soils, Planosols	Kaolinite, halloysite, vermiculite, 2:1/2:2 intergrades, sesquioxides, gibbsite
9. Oxisols	Oxide	Latosols, Laterite soils	Sesquioxides, gibbsite, kaolinite, 2:1/2:2 inter-grades
10. Histosols	Histology	Bog soils	Variable

From Soil Survey Staff (1960). Also from Jackson (1964), in Chemistry of the Soil, ed. F. E. Bear, by permission. © Van Nostrand Reinhold Co., New York.

When water falls on land in the form of either rain or snow, some of it penetrates surficial sediments. Infiltration occurs until the rate of supply of water exceeds the capability of the material to absorb it. A material's *infiltration rate* is the rate at which it absorbs water. Its *infiltration capacity* is the balance between the rate of supply and the rate of infiltration. When its infiltration capacity is exceeded, the material can no longer accept water at the rate of supply, and water must run off the land.

Water available for runoff is part of a balanced system of water added and water withdrawn. The formula for water balance *(Wb)* is:

$$Wb = P - (R+D+E+T)$$

where P is precipitation, or water added; R, runoff; D, percolation to ground water; E, evaporation from the land surface; and T, transpiration from vegetation. The last two are generally combined as ET, or evapotranspiration. Runoff, percolation, and evapotranspiration are water withdrawn.

Runoff requires a sloping land surface, and flow downward is either unconfined (not channeled) as *sheet flow*, or confined as *channel flow*. A stream is a body of water flowing in a channel. To understand flow one must know a few principles of hydraulics.

Elementary hydraulics

From physics we know that fluids are mobile and that movement within them is caused by pressure

3

Runoff and Streams

(a)

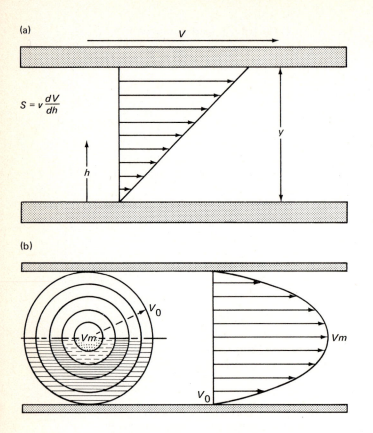

$$S = v \frac{dV}{dh}$$

(b)

Figure 3.1 Laminar flow between plates (a) and through a straight cylindrical pipe (b).

differences between any two points. Movement is displacement (*d*); pressure is applied force (*F*). Their product is work (*W*):

$$W = Fd$$

Natural applied force is gravity, and in a sloping channel the resolved gravitational force is *F* sin *i*, where *i* is the channel gradient. Force is the product of mass (*m*) and acceleration (*a*):

$$F = ma$$

Therefore another formula for work is:

$$W = mad$$

The capacity for doing work is energy, and for a fluid in motion that capacity is kinetic energy (*KE*):

$$KE = \tfrac{1}{2}mV^2$$

Thus, the energy of flowing water, its capacity for doing work, is influenced mainly by the velocity of flow. The work done by a stream is erosion, transportation, deposition.

Fluids have great molecular mobility and when in motion develop intermolecular friction, which creates viscosity. If gravitational force causes fluid motion in a channel, opposing and resisting forces develop within the fluid and at the contact of water with the channel bottom and sides. Examine fluid motion between two parallel plates [fig. 3.1(a)]. The lower plate is at rest while the upper plate, parallel to it, is moving at a uniform rate. Liquid adheres to each plate. The velocity of the liquid at the lower plate is zero. The velocity of the liquid at the upper plate equals the velocity of the plate. The velocity of the liquid at any point between the plates is proportional to the distance (*h*) above the lower plate. To maintain the uniform velocity of the upper plate, an applied tangential force or shearing stress must overcome friction within the fluid. Shearing stress (*S*) is proportional to the velocity gradient between the plates:

$$S = v \frac{dV}{dh}$$

where *v*, viscosity, is a constant of proportionality between shearing stress and the velocity gradient.

Now examine fluid flow through a straight cylindrical pipe of uniform diameter [fig. 3.1(b)]. Because of adhesion, the velocity is zero at the pipe wall. Along the axis of the pipe, the velocity is the maximum. Between the wall and the axis, concentric cylindrical layers move at different velocities and slide past each other. This velocity pattern is like the surface of a projectile point; it is a paraboloid of revolution. Half of a vertical section of the pipe reveals the flow pattern, which approximates the flow pattern in an open channel of similar shape. From the centrally located semicircular layer with maximum velocity, concentric layers extend to the channel wall, each succeeding layer having lower velocity, until at the wall the velocity is zero.

These two cases, in which parallel fluid layers move past one another at different velocities, are examples of *laminar flow*, in which water particles travel along a definite path and never intersect the paths of any other particles. Flow is laminar at low velocities. As velocity increases, eddies form, and water paths are irregular, and they twist, cross, and recross at random in a state of *turbulent flow*. When velocity then decreases, flow becomes laminar again; the point at which this happens, the velocity is lower than it was during the initial laminar flow. This lower limit is the *lower critical velocity,* below which flow is always laminar and is expressed by the *Reynolds number (R):*

$$R = \frac{Vd}{v/n}$$

Where V is velocity; d, depth; v, viscosity; and n, density.

Boundary shear at the wall influences flow velocities throughout the water cross section, and in open channels it is expressed by the *Chezy equation:*

$$V = C\sqrt{RS}$$

where V is the mean velocity; C, the Chezy constant depending on forces contributing to friction; R, is the hydraulic radius; and S, slope. The *hydraulic radius* is the ratio of the cross-sectional area of the channel to its *wetted perimeter (P)* which is the length of the water-channel contact at the cross section [fig. 3.2(a)]. Hydraulic radius is expressed in feet and is approximately equal to mean water depth in wide, shallow channels.

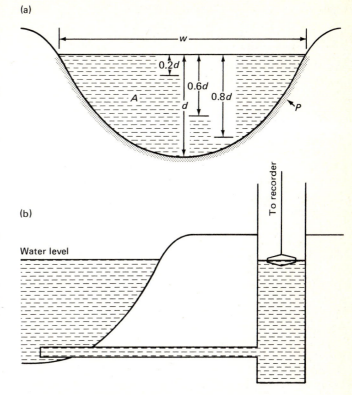

Figure 3.2 (a) A hypothetical cross section of a stream channel: A, cross-sectional area; w, water width; d, depth; P, wetted perimeter. In a two-point method of measuring vertical velocity, velocities are measured at 0.2 and 0.8 foot depths. In a one-point method, velocity is measured at 0.6 foot depth. (b) Stilling well and stream channel.

Channel roughness, which also controls velocity, is evaluated in the *Manning equation:*

$$V = \frac{1.486}{n}\ R^{2/3}\ S^{1/2}$$

where V is the average velocity in feet per second; n, the Manning roughness coefficent; R, the hydraulic radius in feet; and S, the slope of the channel in feet per foot. Values for n have been determined for many kinds of channels and conduits (table 3.1).

Table 3.1 Values for *n* in the Manning equation

Channels or structures	*n*
Excavated ditches and channels	
Earth	
Unlined, straight, uniform	0.020 – 0.025
Unlined, irregular	0.025 – 0.035
With light vegetation	0.035 – 0.045
With fairly heavy vegetation	0.040 – 0.050
Drag-line excavation	0.028 – 0.033
Rock	
Smooth, uniform	0.030 – 0.035
Jagged, irregular	0.035 – 0.045
Lined channels and ditches	
Concrete	0.013 – 0.022
Concrete sides, gravel bottom	0.017 – 0.020
Riprap	0.020 – 0.030
Riprap sides, gravel bottom	0.023 – 0.033
Asphalt	0.013 – 0.016
Natural channels	
Clean straight banks, uniform bottoms, full stage	0.027 – 0.033
Same, with some vegetation	0.033 – 0.040
Clean, meandering, with minor pools and eddies	0.035 – 0.050
Sluggish, winding channel, with deep pools	0.060 – 0.080
Same with dense vegetation	0.100 – 0.200
Rough, rocky, in mountain terrain	0.050 – 0.080
Overflow areas adjacent to regular channel	0.030 – 0.200
Structures	
Concrete pipe	0.013 – 0.015
Clay pipe	0.012 – 0.014
Corrugated metal pipe	0.019 – 0.024

From Woodward and Posey (1941), by permission. © John Wiley & Sons, Inc., New York.

The Manning equation is also used for estimating velocity in channels. A series of typical cross sections are superimposed, and an average section is compiled that gives values for *R* and *S*. The equation is solved using a value for *n* (West, 1960).

The amount of water flowing in a channel is directly dependent on velocity:

$$Q = AV$$

Where Q is discharge, or the volume of water passing through a channel cross section per unit of time, expressed in cubic feet per second; A, the channel cross section in square feet; and V, mean velocity in feet per second. This basic flow equation is used in channel hydrology to evaluate many of the activities of a stream.

Working hydrology

Many streams are gauged in the United States and elsewhere. In this country, daily discharges are routinely reported for each water year, which runs from October 1 through September 30 of the following calendar year. Among other data, total, mean, maximum, and minimum monthly discharges are reported.

Measurement of discharge

To determine discharge, one measures mean velocity and the channel cross section at a station. A *current meter,* which measures stream velocity, consists of cups on a wheel that rotates when struck by flowing water. Electrical contacts count each revolution or every fifth revolution of the wheel. Listening through an earphone, one determines velocity by timing the number of clicks or revolutions with a stopwatch. The cup assembly is moved up or down on a graduated wading rod or by use of a cable. For depths greater than 2.5 feet a larger Price meter is used; for shallower depths, a pygmy meter.

In the field a channel cross section is divided into vertical slices of equal width [fig. 3.2(a)]. Along the vertical center line of each slice, velocity measurements are made at 0.1 depth increments, and at 0.5 foot beneath the water surface and above the stream bed if a Price meter is used (0.3 foot if a pygmy meter is used). A *vertical-velocity curve* is plotted for each slice (fig. 3.3), with depths plotted proportionally to total depth. Mean velocity of the vertical slice is the ratio of the area between the curve and the ordinate axis to the length of the ordinate axis.

In a *two-point method*, velocities are measured at 0.2 and 0.8 depths [fig. 3.2(a)]. The average of these two velocities is the mean velocity of the vertical slice. In a *one-point method*, velocity is measured at 0.6 depth, and results are usually reliable for channel depths of less than 2.5 feet.

Discharge is calculated by summing the slices:

$$Q = \sum (av)$$

where Q is the total discharge; a, the cross-sectional area of a vertical slice; and v, the mean velocity through the vertical slice. For further details about measuring discharge see the instructional manuals of the U.S. Geological Survey (Carter and Davidian, 1968; Buchanan and Somers, 1969), which also discuss measuring discharge by weirs, portable measuring flumes, floats, and dyes.

Discharge may be estimated from the Manning equation. By combining it with the general equation $Q = AV$, the Manning equation may be written:

$$Q = \frac{1.486}{n} \ AR^{2/3} \ S^{1/2}$$

This slope-area method gives reasonable results where direct measurement is impractical or impossible (West, 1960; Dalrymple and Benson, 1967).

An automatic recording device is used to obtain a continuous discharge record. In one standard method, a *stilling well* is dug beside a channel, and a pipe is installed from the channel bottom to the well [fig. 3.2(b)]. Water rises in the well to the level of the water in the channel. The water's rising and falling in the well causes up-and-down movement of a float, and the water level is graphed or digitized automatically by the recorder housed above the stilling well. Other devices also record water levels (Buchanan and Somers, 1968).

Water level is reported according to an established datum such as mean sea level or some arbitrarily selected level. The height of the water surface above the datum is the *stage*. To avoid negative values, the datum is always set below zero flow of the stream at the station.

Water level and discharge are related in a *rating curve* for the gauging station. At a given stage, discharge is determined by current-meter measurements of vertical slices. After many paired observations, a curve is computed and plotted. The discharge then may be calculated from stage.

All changes in water level must be due to the rise or fall of the water alone. If the channel bottom rises or falls the depth will be affected. To avoid this problem, one selects a stable cross section in bedrock or places a control structure in the channel to maintain width and depth at constant values.

Hydrographs

A *hydrograph* is a continuous graphic record at a gauging station that shows the discharge past the gauge through time. A commonly used graph shows daily discharge throughout a water year. A smooth, sinuous curve drawn through the graph at the base of prominent peaks gives an estimate of *base flow*, which is the discharge supplied by ground water to the channel. Actually, base flow may be somewhat above this generalized curve in any of the peaks (Kunkle, 1968).

Peak discharges represent responses to some

hydraulic input. In the sample hydrograph (fig. 3.4), note the small supply of water (snow) from early December to late March. A rise in ground water (water-level) in the well during March was followed by a rise in water level in the channel in early April. Peak discharge resulted mainly from increased base flow.

The result of many storms during early June was a peak discharge during the latter part of the month, at which time the ground-water reservoir was being depleted. However, storms later in June and in July recharged the reservoir, and the ground-water surface rose in late July. Peak discharge resulted mainly from *runoff* rather than from base flow.

These examples show the complicated response of channel flow to rainfall and base flow.

Note the precedence, contemporaneity, and lag in the different parts of the system.

The measurement of discharge is uncomplicated where water is confined to a channel, but if water spills over banks during floods, other measurement procedures must be used (Matthai, 1967). After peak discharges of the flood are determined, a flood frequency analysis is made.

Flood frequency analysis

Stream hydrographs are used to analyze the magnitude and frequency of floods at a given station. All floods are listed, with the flood of greatest discharge as number one, the flood of second greatest discharge as number two, and so on. Since there may be more than one flood of a given

Figure 3.3 Typical vertical-velocity curves beneath free-water surface (left) and under ice (right). From Buchanan and Somers (1969).

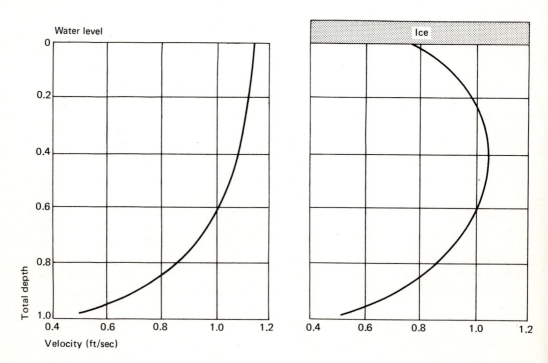

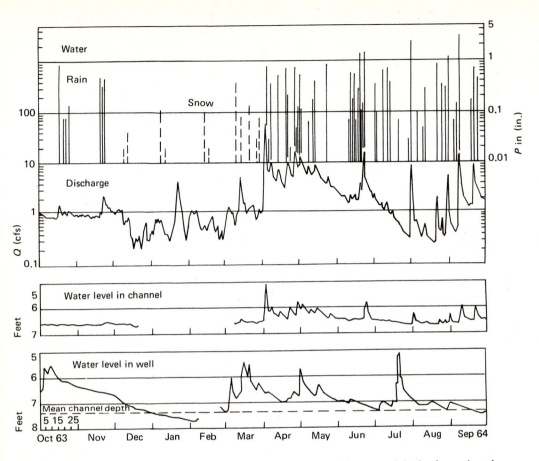

Figure 3.4 A stream hydrograph showing discharge for a water year in relation to precipitation from rain and snow, water level in the channel, and water level in flood-plain well.

discharge, a flood recurrence interval (*RI*) is calculated:

$$RI = \frac{N + 1}{M}$$

where *N* is the number of years that a given flood recurs and *M* is the number that is assigned in listing flood discharges.

The recurrence interval is the number of years, on the average, until a given flood will be equaled or exceeded one time. It is inversely related to the chance that a given flood will be equaled or exceeded in any one year. (A 20-year flood would have one chance in 20, or a 50-year flood one chance in 50, of being equaled or exceeded in any one year, and so forth.)

A flood-frequency curve can be constructed by plotting discharge against the recurrence interval (fig. 3.5). The magnitude of a given flood and its probability of recurrence may be predicted from the graph. In many places, however, discharge records are available for only three or four decades, so values for 50-year or 100-year floods can be estimated only by extrapolation. Prediction

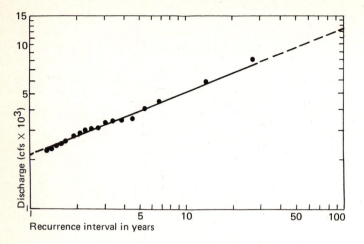

Figure 3.5 A flood-frequency curve for the Skunk River near Ames, Iowa. Data from Schwob (1953).

beyond measured data must be tempered with caution.

A flood plain is the part of a river valley that is subject to flooding. The extent to which the flood plain is covered by floods of different magnitudes should be known. Mapping these flood-prone areas requires an analysis of the topography, the surface and subsurface drainage, the effects of man-made obstructions, and other features.

Hydraulic geometry of channels

On the flood plain, stream channels have hydraulic characteristics of depth, width, and velocity that vary with discharge as simple power functions: $Y = aX^b$ (Leopold and Maddock, 1953). At a channel cross section:

$$w = aQ^b \quad d = cQ^f \quad V = kQ^m$$

where w is water surface width; d, mean water depth; and V, mean velocity. These parameters are plotted against discharge (Q) as a straight line on log-log graph paper. For least squares fit of the power function, the equation used is:

$$\log Y = \log a + b \log X$$

and the constants $\log a$ and b are determined.

That hydraulic parameters are interdependent is shown by combining the power functions with the basic flow equation

$$Q = AV \quad \text{or} \quad Q = wd\, V$$

$$Q = aQ^b \times cQ^f \times kQ^m \quad \text{or:}$$

$$Q = ackQ^{b+f+m}$$

Both the product of the constants and the sum of the exponents equal one.

Discharge increases downstream because drainage area increases. Width, depth, and velocity also tend to increase (Leopold and Maddock, 1953). At any place downstream, the values of the variables, constants, and exponents change, but the interaction that takes place maintains the adjustment in the interdependent system. In some streams velocity does not increase downstream, however (Carlston, 1969). In the Missouri, Potomac, and Susquehanna Rivers, velocity is essentially constant downstream, although discharge increases. In very large streams increase in discharge downstream is compensated for mainly by increase in depth; in smaller streams, by increase in width. Thus, width, depth, and velocity change with discharge at a cross section and progressively change downstream. Adjustment within the interdependent system, which is characteristic of most natural channels, demonstrates the establishment of a quasiequilibrium (Leopold and Maddock, 1953).

Adjustment in a channel in unconsolidated material is shown by concurrent changes in mean water depth and mean channel depth. At 10 closely spaced cross sections along Four Mile

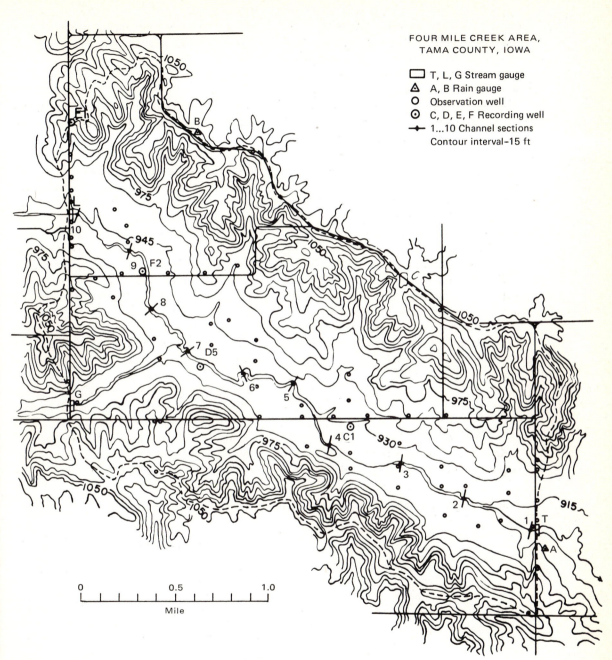

Figure 3.6 Field measurement stations along Four Mile Creek, Iowa.

Creek in Iowa (fig. 3.6), channel depth and water depth were measured at successive time intervals. At a cross section a decrease in channel depth ($Dc-$) indicated fill; an increase in channel depth ($Dc+$), scour. Concurrent water-depth measurements showed that scouring or filling occurred regardless of the rise or fall of the water. In addition, the deepest part of the channel, the *thalweg,* shifted to the left or right regardless of whether the water rose or fell or remained constant.

Scour and fill in this channel were related to change in hydraulic energy expressed as mean velocity in feet per second (fig. 3.7). As mean velocity increased ($Vc+$), filling occurred ($Dc-$), which was followed by scouring ($Dc+$) as velocity decreased ($Vc-$). This cyclic behavior of filling and scouring occurs as well in large streams such as the San Juan River in Utah and the Colorado River in Arizona (Leopold and Maddock, 1953). There, filling during periods of increasing velocity has been attributed to the concurrent increase in suspended load.

The work of a stream

As velocity is a dominant factor in channel hydraulics and stream energy, the work done by a stream – erosion, transportation, and deposition –is often related to it.

Erosion

Running water erodes by hydraulic impact, corrasion, and corrosion. *Hydraulic impact* removes material by the force of the water alone. When a garden hose is jetted on barren soil, a hole is formed. A plunge pool is also formed at the base of a scarp where freely falling water impacts material. A cubic foot of water dropping vertically 64 feet has 4000 foot-pounds of energy at impact on a square foot at the base of the wall.

Corrasion is the solution of soluble material as water flows across it. *Corrasion* is the process whereby particles in transport grind and abrade rock and other particles. The effect is like that of a ball mill.

Erosion creates the total load that a stream carries—dissolved solids, suspended load, and bed load. *Dissolved solids* are the chemical elements and compounds that are carried in solution. They are provided by corrosion in the stream channel, by dissolved solids in base flow from ground water, by surface runoff to the stream, and by supply of aerosols from the atmosphere. The total dissolved solids (TDS) contain many cations, of which Ca^{++}, Mg^{++}, Na^+, and K^+ are

Table 3.2 Analyses of waters of Wabasha County, Minnesota (ppm)

Constituents	Streams			Ground water				Jordan sandstone
				Surficial deposits				
	(1)	(2)	(3)	(4)	(5)	(6)	(7)	(8)
SiO_2	7.9	8.9	2	8	9.7	2.7	4.6	20
Ca	49	56	96	96	76	81	61	57
Mg	15	21	29	35	27	27	28	26
Na + K	–	8.9	–	22	11	15	10	3
HCO_3	213	293	394	468	358	286	289	289
SO_4	9.9	4.8	26	20	29	35	48	16
Cl	–	1.2	6.2	20	2.6	11	3.6	1.7

From Thiel (1944), by permission of Minn. Geol. Surv.

usually the most abundant, as one would expect, since base-forming cations are released in weathering. The dominant anions are usually HCO_3^-, $SO_4^=$, Cl^-, and NO_3^-. Heavy metals, phosphates, and phenols are not uncommon (Taras et al., 1971). The dissolved solids in a stream are generally related in kind and abundance to those contained in ground water in surficial deposits and bedrock of the surrounding area (table 3.2).

Surface water crossing carbonate terrain has appreciable amounts of Ca^{++} and HCO_3^- and is generally neutral to slightly alkaline. Lost River crosses a karst area in south-central Indiana. Upstream, concentrations of Ca^{++} and Mg^{++} are 58 to 60 parts per million (ppm) and 8 to 10 ppm, respectively; downstream about 9 miles, 75 to 100 and 14 to 17 ppm, respectively. Alkalinity, reported as ppm $CaCO_3$ equivalent, increases from 160 to 220. *Alkalinity* is the acid-neutralizing capacity of the water, or the amount of various alkalis in the water capable of neutralizing acids. The alkalis are dissolved bicarbonates, carbonates, and hydroxides of Ca, Mg, and Na. Upstream, pH is 7.7 and 7.8, while downstream it is 6.9. It decreases because $SO_4^=$ increases from 10 to 12 ppm upstream to more than 40 ppm downstream.

Lost River has hard water. *Total hardness* is the concentration of Ca^{++} and Mg^{++} ions expressed as $CaCO_3$ equivalent ppm. In Lost River total hardness increases from 150 to 220 ppm upstream to more than 300 ppm downstream.

In areas near a seacoast, such as San Mateo County, California, Na^+ and Cl^- are carried in atmospheric moisture from the sea and are precipitated inland. Soils unaffected by fallout have Na^+ and Cl^- saturation extracts of less than 1 me/l, but *B* horizons of affected soils have Na^+ extracts of 10 to 21 me/l and Cl^- extracts of 10 to 28 me/l.

In the atmosphere nationwide (Junge and Werby, 1958), belts of higher concentration of Na^+ and Cl^- parallel the Pacific and Atlantic coasts and the Gulf of Mexico. There is a concen-

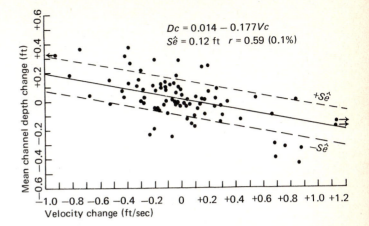

$$Dc = 0.014 - 0.177Vc$$
$$S\hat{e} = 0.12 \text{ ft} \quad r = 0.59 \ (0.1\%)$$

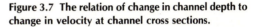

Figure 3.7 **The relation of change in channel depth to change in velocity at channel cross sections.**

tration of Ca^{++} over the desert regions of the Southwest and the Great Plains, where dust storms are common and where many soils have prominent *ca* horizons. When carbonate dust dissolves in atmospheric moisture it releases Ca^{++} ions in aerosols.

In fallout, aerosols introduce dissolved solids directly to streams and to soils. Runoff from soils to streams and percolation to ground water and base flow add dissolved solids to the total stream load.

Transport of sediment

In addition to its dissolved load, a stream has a *suspended load*, which includes the particles that are moved in suspension. Other particles may traject into the flowing water, but shortly thereafter they fall to the bed with a movement like a series of bounds, termed *saltation*. Other particles, which roll and slide along the bed, are part of the *bed load*. Suspension and saltation of particles depend on the stream's turbulence and the particles' settling velocity.

The removal of a particle from the surface of the bed depends on hydrodynamic forces exerted on the particle by the fluid in motion, and on the particle shape, its immersed weight, its distance of protrusion above bed level, its external bearing points relative to other particles, and its possible impact by other moving particles causing dislodgment. The hydrodynamic forces are critical tractive force and critical velocity. *Critical tractive force* is the ratio of the effective force of water paralleling the bed to the resistance of a particle on the bed. It is a function of the ratio of particle size to the thickness of the laminar boundary layer. *Critical velocity* is a fluid-driving force that is proportional to the square of the local flow velocity and the projected area of the particle (Chen, 1970).

Examine forces relative to a particle protruding above the general level of the other particles of a bed (fig. 3.8). Fluid approaches a particle with a velocity differential (Va). Velocity is greater at the top of a particle than at the bottom. Fluid driving force (F) is the resultant of shear stress and pressure over the surface of the particle. The resultant force resolves into drag force (D) in the direction of flow and lift force (L), which is normal to the direction of flow. The particle tends to rotate as in a couple (C), and the rotation is transmitted

Figure 3.8 Forces exerted on a particle protruding from the average bed level in a channel. From Chen (1970).

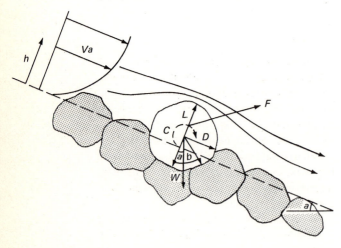

through an angle of repose b to a bearing point on an adjacent particle. If the resultant (F) is greater than the opposing forces including immersed weight, the particle is moved.

The bed load shifts as flow regimes change. Tranquil and rapid flow are expressed by the *Froude number* (F) in the relation:

$$F = \frac{V}{\sqrt{gD}}$$

where V is mean velocity; g, gravitational force; and D, water depth. If flow is tranquil then F < 1, but if flow is rapid F > 1. In the lower flow regime, bed configuration passes through three phases (fig. 3.9). In phase 1 the bed is smooth, with no sediment movement and little turbulence. In phase 2 ripples form on the bed and there is little sediment movement and little turbulence. In phase 3 large irregular dunes form on the bed, sediment moves downstream, and flow is turbulent.

As flow passes into the upper regime in phase 4, dunes are smoothed to a plane bed, and the water surface is smooth with little turbulence. In phase 5 smooth sinusoidal waves form in a fixed position on the bed. Surface waves, which are also sinusoidal, are in phase with the bed waves. In phase 6 symmetrical sinusoidal waves form on the bed and progress upstream with increasing amplitude, then collapse and re-form. Symmetrical surface waves progress upstream in phase with bed waves and increasing amplitude until the waves break. The whole system then collapses and re-forms (Simons and Richardson, 1966).

Flow and bed phenomena are related to stream power, *VfDS*, where V is average velocity in feet per second; f, the specific weight of the water-sediment mix in pounds per cubic foot; D, the average depth of flow in feet; and S, the slope of the energy-gradient line, which is equal to the water-surface slope in uniform flow (Simons and Richardson, 1966). The stream-power factors f, D, and S are part of the *DuBoys equation*, which is generally used to evaluate tractive force in a stream:

$$T = fDS$$

where T is the tractive force; f, the specific weight of water; D, the depth of water; and S, the stream gradient.

Since hydraulic radius (R) is about equal to D, stream power is expressed as $62RSV$ in foot-pounds per second per square foot (Benson and Dalrymple, 1967) and is related to the particle size of the bed load within flow regimes of a stream (fig. 3.10). Where the median particle size of the bed load is small and stream power is low, flow regimes shift from dune to antidune phases through a low but narrow range of stream power. Where the median particle size of the bed load is large, a greater and wider range of stream power is needed to shift from lower to upper flow regime.

Critical velocity is the lowest velocity required for moving a loose grain of a given size on the bed of a channel. The importance of critical velocity for movement and transport of particles is shown in the famed *Hjulström diagram* (fig. 3.11). At a velocity of 15 to 17 cm/sec, particles of fine and medium sand (0.1 to 0.4 mm) are moved. Greater velocities are required for moving coarser sand, pebbles, and cobbles. A cobble with a diameter of 70 mm has a critical velocity of 270 cm/sec. But very fine particles (less than 0.1 mm) also have a greater critical velocity. Velocities of 46 to 76 cm/sec are needed to move medium silt (0.016 to 0.008 mm), velocities of 200 to 260 cm/sec for clay (less than 0.002mm). Cohesion among finer particles causes resistance to movement, which makes higher velocities necessary.

The Hjulström diagram also shows that transport of pebbles and cobbles (greater than 2 mm) is maintained only through a narrow range of velocities. Very fine silt and clay (less than 0.004 mm) remain in transport through a wide range of velocities. Velocity requirements separate particles into bed loads and suspended loads.

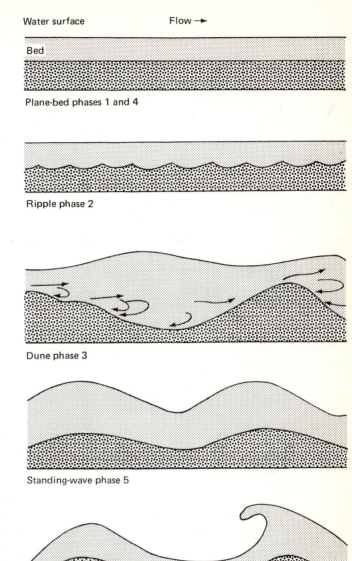

Plane-bed phases 1 and 4

Ripple phase 2

Dune phase 3

Standing-wave phase 5

Antidune phase 6

Figure 3.9 Flow regimes and bed configuration in a channel. From Simons and Richardson (1966).

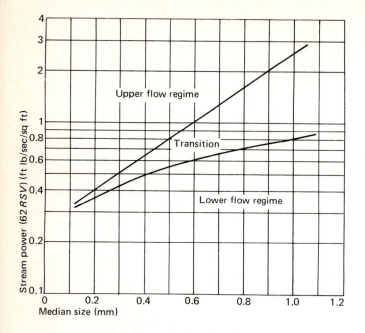

Figure 3.10 The relation of stream power to particle size and flow regimes. From Benson and Dalrymple (1967).

If the suspended load is plotted against discharge on log-log graph paper, a straight line may be fitted through the scatter diagram. The relation may be expressed as (Leopold and Maddock, 1953):

$$L = pQ^j$$

where L is suspended load in tons per day; Q, discharge in cubic feet per second; and p and j, constants. Suspended load increases with discharge, but it increases at a more rapid rate. As suspended load and velocity are both related to discharge, velocity must be a controlling factor of suspended load.

Deposition

Increase in velocity initiates transport, and decrease in velocity stops transport and causes deposition. A cobble with a diameter of 70 mm (fig. 3.11), whose movement starts at a stream velocity of 270 cm/sec, remains in transport until stream velocity decreases to 180 cm/sec, when transport ceases and deposition occurs. A very fine sand particle (0.1 mm), whose movement starts at a stream velocity of 17 cm/sec, remains in transport until stream velocity decreases to about 0.7 cm/sec, when transport ceases and deposition occurs. A velocity decrease of only 33 percent is required for deposition of the cobble, a decrease of 95 percent for the sand.

Velocity differentials cause sorting of sediment in a channel according to particle size. Pebbles (4 mm) are deposited when velocity decreases to about 30 cm/sec, but at this velocity sand and silt particles (2 to 0.62 mm) erode and are transported.

Very fine particles (less than 0.01 mm) remain in transport or suspension at velocities of less than 0.1 cm/sec, but they settle in quiescent water, where *Stokes' law* applies. The resistance that a fluid offers to a spherical particle settling through it depends on the surface area of the particle ($6\pi r$), the viscosity of the fluid (v), and the velocity of the fall (V). The buoyant force tending to hold the particle suspended depends on the volume of the particle sphere ($4/3\ \pi r^3$), the density of the fluid (d_2), and the gravitational force (g). The buoyant force is balanced by a force tending to pull the grain down, which is dependent upon the volume of the particle sphere, the gravitational force, and the density of the particle (d_1). The net force acting downward is:

$$\frac{4}{3}\ \pi\ r^3\ (d_1 - d_2)\ g$$

When resistance equals net force, velocity (V) becomes constant and remains so, or:

$$6\ \pi\ rvV = \frac{4}{3}\ \pi r^3\ (d_1 - d_2)\ g$$

Thus, Stokes' law is expressed as:

$$V = \frac{2}{9}\ \frac{(d_1 - d_2)\ gr^2}{v}$$

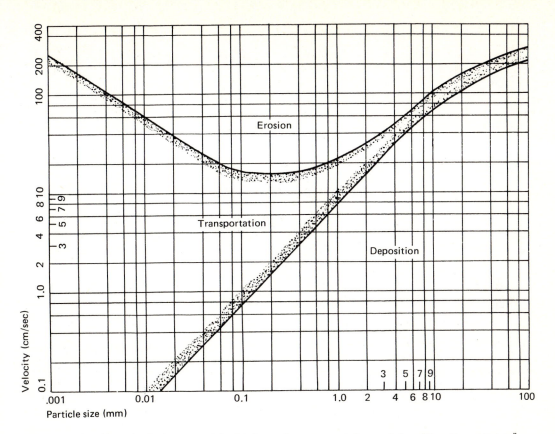

Figure 3.11 The effect of velocity on particle size in erosion, transportation, and deposition. From Hjulström (1935).

which applies only to silt and smaller-size particles. If all particles have the same density, the larger particles fall through a given distance at a faster rate than the smaller particles, which creates vertical sorting and bedding.

Channel shape

Channel shape is described by cross-sectional dimensions, longitudinal profile, and channel pattern.

Longitudinal profile

The longitudinal profile of a stream is described by its continuous fall in elevation and the hori-

zontal distance between source and mouth. For most streams (Stall and Yang, 1970), the profile is a curve that is concave upward, steeper near the source but gradually flattening toward the mouth. The curve may be described by a mathematical equation, and at any point, channel slope is a tangent to the curve.

Along the longitudinal profile, channel slope, stream length, drainage area, discharge, and bed-load characteristics are interrelated (Hack, 1957). Average annual discharge in cubic feet per second (Y) increases as drainage area in square miles (X) increases. This relationship is expressed by a power function, $Y = aX^b$.

Channel slope similarly relates to drainage area, but slope decreases as area increases. The

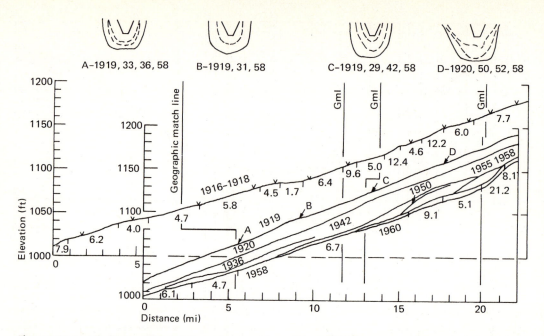

Figure 3.12 Longitudinal profiles and channel cross sections of the Willow River, Iowa. Slopes in feet per mile are given between ticks. The junction of tributaries is marked by ν. Comparable geographical points are matched in profile sets. From Daniels (1960), by permission of Am. J. Sci.

rate of change of slope depends on the kind of rock the stream crosses. Geological control is exerted on stream slope by different kinds of rock or material. Slope is also related to the median particle size of the bed material, since rates of change of slope correspond to different sizes of bed-load particles. Slope increases as a function of the particle size of the bed-load.

In streams whose bed material is of equal-size particles, channel slope is inversely proportional to drainage area or discharge. Where drainage areas are equal, channel slope is directly proportional to the particle size of the bed material. In areas having similar geological characteristics and drainage areas, streams have similar channel slopes and bed material of similar particle sizes.

Stream length is the distance from a point on a stream to the drainage divide at the head of the longest stream above the reference point. Length (L) is related to drainage area (A) as expressed by a power function, $L = aA^b$. Median size of bed particles in mm (M) is directly related to stream length in miles (L) as $M = jL^m$.

As stream slope (S) and stream length (L) are related to other common parameters, they are related to each other, as expressed by $S = kL^m$. This relation between slope and length and their relation to bedrock, bed load, drainage area, and discharge indicate that the longitudinal profiles are adjusted to the work that must be done. This adjustment suggests that an equilibrium exists (Hack, 1957). The longitudinal profile of a stream, where these adjusted conditions exist, has been termed the "profile of equilibrium" for some time.

A stream is considered to be in a quasi-equilibrium state when the total rate of work done, with a given discharge and a given maximum relief, decreases as profile concavity increases (Langbein and Leopold, 1964). Discharge increases downstream because of greater drainage area, but concavity increases upstream in the headwaters, where discharge is small. Width, depth, velocity, slope, and sediment-transporting ability adjust to excessive concavity or straightening of the profile.

A stream in "equilibrium" or "quasi-equilibrium" is not unlike a *graded stream* (a term that has been in the literature for years), that is, a stream in which over a period of years, slope is adjusted to provide, with available discharge and with prevailing channel characteristics, just the velocity required for transport of the load (Mackin, 1948).

The longitudinal profile of a stream may be interrupted at many places by distinct changes in gradient caused by waterfalls, rapids, knickpoints, or mounds of sediment in the channel. The natural, sinuous channel of the Willow River in western Iowa flows in silty alluvium. In 1916 its width was 60 to 100 feet, and its depth was 10 to 12 feet. The average slope was 5.2 feet per mile (Daniels, 1960). Where major tributaries joined the main stream, slope flattened because of the sediment mounds built in the main channel by the tributaries (fig. 3.12).

Adjustments were noted following man-made changes in hydraulic factors (Ruhe, 1971). To prevent flooding the stream was ditched with bottom widths of 12, 10 and 8 feet progressively upstream. The angle of the ditch was 45° with a berm width of 15 feet. The ditch cut across channel bends, and stream length was shortened from 28 to 22 miles. Slope increased from 5.2 to 7.7, 8.5, and 12.1 feet per mile (Daniels, 1960). The ditch was constructed in a lower reach from 1916 to 1919, but stream reaction was so rapid that ditching had to be continued in an upper reach during 1919 and 1920.

Artificial changes in length and slope caused immediate deepening and widening (Daniels, 1960). At four measured cross sections (A through D) width increased 2.5 to 3.5 times the constructed width, and depth increased 1.3 to 3.0

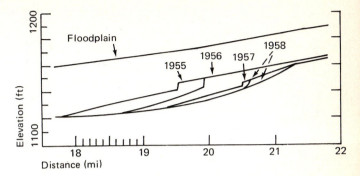

Figure 3.13 Advance of knickpoint in the Willow River from 1955 to 1958. From Daniels (1960), by permission of Am. J. Sci.

times the constructed depth. These changes occurred in less than 40 years. Throughout 15 miles of the lower reach, slope flattened, and gradients are now 4.7 to 6.7 feet per mile, and they almost parallel the natural channel slope prior to ditching.

Adjustments continue in the upper reach by headward migration of knickpoints in the channel. Knickpoint movement is caused by a waterfall creating a plunge pool at the downstream base of the scarp. Swirling water and wave action undercut the scarp. Shear planes form in undercut sediment, and blocks slump into the pool. A vertical face is re-established, and the process is repeated. Knickpoint movement is not continuous or uniform. One knickpoint moved one mile upstream during the years 1953 to 1956 and another half mile in 1956 and 1957 (fig. 3.13). Between July, 1957, and April, 1958, it moved only 90 feet, increasing in height from 1.5 to 3 feet. During April and May, 1958, it moved 600 feet but then remained fixed until July, 1958. During the next two months it moved 1400 feet upstream.

In the Willow River, the decrease in length caused an increase in slope, which in turn affected discharge and velocity. Width and depth increased, and slope flattened as the stream ad-

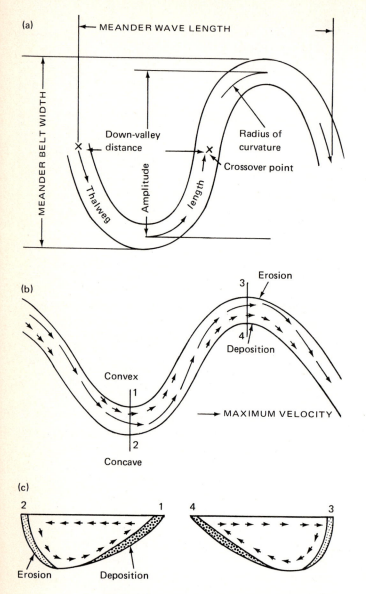

(a)

MEANDER WAVE LENGTH

MEANDER BELT WIDTH

Down-valley distance

Radius of curvature

Crossover point

Thalweg

Amplitude

length

(b)

Erosion

3

4

Deposition

Convex

1

MAXIMUM VELOCITY

2

Concave

(c)

2   1   4   3

Erosion   Deposition

Figure 3.14 The geometry and flow paths of stream meanders and bends. From Leopold and Wolman (1960), by permission of Geol. Soc. Am.

justed. The channel is relatively straight and has not returned to its original sinuous pattern.

Channel pattern

Channel pattern is the planimetric or map view of a stream. Patterns are straight, sinuous, or braided. *Sinuosity (U)* is the ratio of the thalweg length *(Lt)* to the down-valley distance *(Dv)* between points of measurements, or:

$$U = \frac{Lt}{Dv}$$

Consider a circle whose diameter is one. The circumference of the circle is πd, or 3.1416 x 1. The diameter *(d)* subtends half the circumference, or 1.57. Let half the circumference represent thalweg distance, and let the diameter represent the down-valley distance. Then, 1.57/1.00 = 1.57, which is the sinuosity of the curve or bend in the channel. A curve or bend is a *meander*. A stream is considered to be meandering where the sinuosity is 1.5 or more. Meanders are in alluvium. If the stream is entrenched in bedrock or other material, such a bend is an *entrenched meander*.

Meanders are described by wave length, amplitude, and radius of curvature (fig. 3.14). Where one bend becomes another is the crossover point. Down-valley distance and thalweg distance are measured between successive crossover points. Most river bends have proportional sizes with consistent relations between wave length, channel width, and radius of curvature (Leopold and Wolman, 1960). Wave length and meander-belt width relate to mean annual discharge as a power function (Carlston, 1965; Schumm, 1967a).

Sinuosity relates to the kind of sediment at the perimeter of the channel (Schumm, 1963). It also depends on other factors, as is shown by the Missouri River from Sioux City, Iowa, to Kansas City, Missouri. The stream was channeled during the 1930s and has been controlled since that time. In its natural condition prior to man's work, the

Table 3.3 Missouri River: length, low-water slope, and sinuosity ratios in 1890 and 1946

Stream reach		Stream miles		Slope (ft / mi)		Sinuosity ratio	
From	To	1890	1946	1890	1946	1890	1946
Sioux City, Iowa	Onawa, Iowa	66.0	45.8		0.98	1.91	1.33
Onawa, Iowa	Crescent City, Iowa	69.0	71.0	0.80*	0.82	1.49	1.44
Crescent City, Iowa	Plattsmouth, Nebr.	32.0	29.0		0.58	1.45	1.32
Plattsmouth, Nebr.	Nebraska City, Nebr.	25.9*		1.23*		1.28	1.28
			43.0		1.24		
Nebraska City, Nebr.	Hamburg, Iowa			0.97*			
Hamburg, Iowa	Rulo, Nebr.	66.0	60.0		1.03	1.48	1.31
Rulo, Nebr.	Kansas City, Mo.	146.8*		0.83*			

*1890 data from Straub and Miller (1935). All other data from Glenn et al. (1960).
Note: Natural channel in 1890; constructed and controlled channel in 1946.
Stream distance from Sioux City, Iowa, to Rulo, Nebraska, was 276 miles in 1890 and 248.8 miles in 1946, a shortening of 27.2 miles.

character of the river changed at the mouth of the tributary Platte River (Straub and Miller, 1935; Whipple, 1942). Above the mouth of the Platte, the Missouri River channel was narrower and more sinuous, its slope was flatter, and there were fewer islands and channel bars. Compare the natural channel of 1890 and the controlled channel of 1946 above and below Plattsmouth, Nebraska (table 3.3). In 1890 the Missouri River meandered 167 miles from Sioux City to Plattsmouth, with sinuosities of 1.45 to 1.91. In the next 43 miles, to Hamburg, Iowa, the stream was sinuous but not meandering, as is shown by the ratio of 1.28. Southward 66 miles to Rulo, Nebraska, the river again meandered, with a ratio of 1.48. Slopes also changed: they were flatter above Plattsmouth, steeper downstream to Nebraska City, and again flatter farther downstream to Rulo and Kansas City.

These channel differences were caused by the effect of the Platte River on the Missouri River. The Platte transports large suspended and bed load and a coarser textured load than the Missouri. From July 1, 1929, to June 30, 1930, the Platte carried a suspended load of 16.8 million tons during a discharge of 5 million acre-feet of water. During the same period the Missouri River above Plattsmouth at Omaha carried 113 million tons in suspension during an annual discharge of 19.9 million acre-feet of water.

The ratios of suspended load to discharge in tons per acre-feet were 3.4 for the Platte and 5.7 for the Missouri. The suspended load of the Platte was about 60 percent of the suspended load of the Missouri. In the following year the suspended load of the Platte was 62.5 percent of the suspended load of the Missouri. In 1929 the bed load of the Platte was 7.6 million tons, or 64.4 percent of the bed load of the Missouri, which was 11.8 million tons. In 1930 these respective values were 7 million tons (83 percent) and 8.4 million tons, and in 1931 they were 3.6 million tons (87.8 percent) and 4.1 million tons (Straub and Miller, 1935).

The mean particle size of the bed load differed, being 0.42 mm in the Platte River and 0.19 mm in the Missouri River at Omaha. About 25 percent of the bed load of the Platte had particles greater than 0.78 mm, compared to 25 percent greater than 0.22 mm for the Missouri (Straub and Miller, 1935).

At the mouth of the Platte River, the Missouri

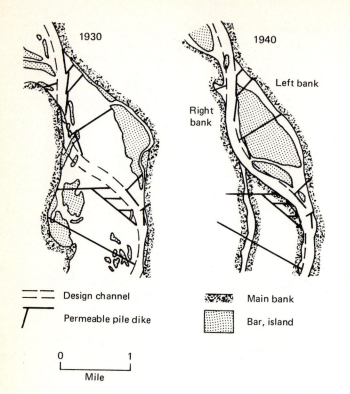

1930

1940

Left bank

Right bank

- - - Design channel

⌐ Permeable pile dike

░░░ Main bank

▦ Bar, island

0 ——————— 1
Mile

Figure 3.15 Otoe Bend, on the Missouri River, Fremont County, Iowa, and Otoe County, Nebraska, showing the natural channel in 1930, the controlled channel in 1940, the design channel, and control structures.

River had to adjust suddenly to an addition of suspended load of 60 to 65 percent already in transport, an addition of bed load of 65 to 88 percent already in transport, and a coarser textured load. Debris dumped in the Missouri River by the Platte created a mound. Slopes flattened upstream and steepened downstream (table 3.3). Sinuosity decreased, channel width increased, and bars and islands formed in the main channel downstream. Because of the increase in suspended load, bed load, and particle size, the Missouri River changed from a meandering stream to a sinuous, braided stream. Adjustments occurred

through 43 river miles, after which the river reverted to meandering (table 3.3). The tributary Platte River, whose drainage area is 90,200 square miles, thus greatly affects the much larger Missouri River, whose drainage area (above the mouth of the Platte) is 323,500 square miles.

In a meander (fig. 3.14), velocity is greater near the concave bank than near the convex bank, which is called a *point*. Current flow generally parallels the water surface from the convex to the concave side, descends, and returns to the convex side along the channel bottom. As velocity is greater on the concave side, the bank erodes. Deposition, which occurs on the convex (or point) side, forms a *point bar*.

These principles were used by engineers for a controlled channeling of the Missouri River (Ruhe, 1971). The engineering design required a generally contracted channel in a series of gentle curves that would change a succession of straight, wide reaches, and involved meanders or abrupt changes in direction that had previously existed (Whipple, 1942). Where the river had to be moved from a given bank, permeable pile dikes were constructed that led away from the bank.

At Otoe Bend (between Fremont County, Iowa, and Otoe County, Nebraska) the channel had to be shifted away from the left and right banks in a wide reach (fig. 3.15, year 1930). Pile dikes were driven in 1934 and 1935 to reduce velocities in the areas selected for deposition and to increase velocities in the areas selected for erosion. As water passed through the dikes, velocity decreased, and sediment load was deposited. As water was diverted around the dikes, velocities increased, and bed and bank were eroded.

When the channel was eroded to the designed alignment on the concave side, the bank was revetted and stabilized. As sediment was deposited behind the dikes, willows were permitted to grow, stabilizing the new deposits. In addition to dikes and revetments, pilot channels were dredged through bars and islands to make cutoffs. In five to six years the channel was essentially in its designed position (fig. 3.15, year 1940). Today the Missouri River is confined to its constructed channel, and all of the land created has been under cultivation for several decades.

In a meander, maximum velocity impacts the concave bank slightly downstream [fig. 3.14(b)], which causes the meander to migrate down valley (Friedkin, 1945). If a chronological sequence of maps and air photographs is available, successive channel positions down valley show the migration. A meander of the Des Moines River southeast of Des Moines, Iowa, moved about a quarter of a mile between 1899 and 1967 (Handy, 1972).

An upstream bend may move down valley faster than the next down stream meander. A narrow neck forms between the bends. During flooding, water spills across the neck, which causes a *neck cutoff* and an *abandoned meander*. Overflow water also may follow a former channel or chute, which causes a *chute cutoff* and an *abandoned channel*.

If there are many bars and islands in a channel, the stream is *braided*. A braided stream flows in two or more channels around alluvial bars and islands. Braided streams are common in semiarid regions, where a large sediment load is supplied to the channel, and in melt-water streams from glaciers.

Braiding occurs in channel reaches where the slope may be oversteepened. In general, braided and straight channels relate to steeper slopes,

Figure 3.16 The braiding of the Missouri River at Otoe Bend, showing the natural channel in 1923, the man-enhanced braiding in 1936 to 1939, and the final complex mosaic.

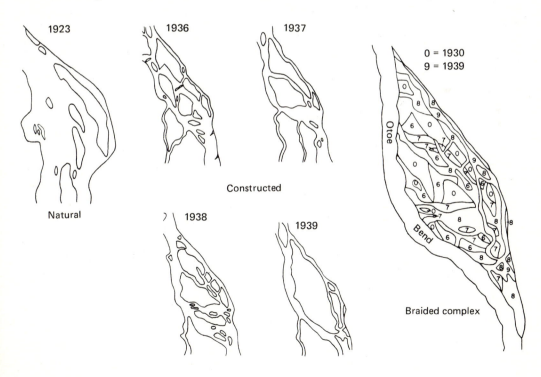

meandering channels to gentler slopes (Leopold and Wolman, 1957). Compare the Missouri River above and below the mouth of the Platte River (table 3.3).

Braiding is also caused by the inability of a stream to carry the total load supplied it. If a local change in velocity occurs, sediment is deposited as bank or bed accretion. This principle was used by engineers at Otoe Bend to control the Missouri River. The placement of permeable pile dikes in 1934 and 1935 (fig. 3.15) purposely reduced velocities in areas selected for deposition and forced the stream to braid. Intricate patterns of bars, islands, channels, and chutes occurred for the next five to six years (fig. 3.16). The final pattern of bank and bed accretion was a complex mosaic of sediments that emerged from the channel at specific times, and after they emerged, these areas were covered only by overbank flood waters.

Not only are many of these features common in stream channels, but they also are preserved as relicts on flood plains and terraces in valleys. They become many of the land forms in the valley landscape.

The major valley features are floodplains, terraces, and alluvial fans, which are stream-constructed landforms. They form by deposition of alluvial sediments and may be shaped by erosional processes.

Flood plains

The *flood plain* is the part of the valley floor adjacent to the stream which is built of sediments by activity of the stream and which is covered with water when the stream overflows its banks. When a stream flows in its channel at bankfull stage, it has a specific discharge, cross-sectional area (or width and mean depth), and velocity. At the instant of bank spillover, a radical change occurs in the basic flow equation of $Q = AV = wdV$. Width increases greatly and rapidly, and mean depth increases slightly and gradually. Velocity adjusts rapidly by decreasing greatly. Transport capability decreases, and deposition occurs. The load is dropped near the bank and then farther from the bank as water flows away from the channel. Water may pond in low areas or become quiescent, and particles may settle, following Stokes' law. The nature of the sediments changes away from the channel.

Valley sediments

The principal types of alluvial sediments are channel, channel-margin, and overbank deposits (Vanoni, 1971). Channel sediments (fig. 4.1) include *transitory channel deposits*, which are

4

Alluvial Landforms

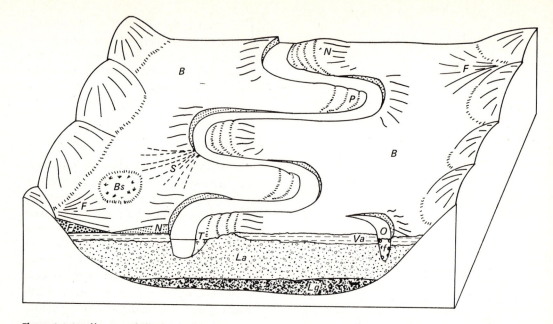

Figure 4.1 Landforms and alluvial deposits in a valley: B, backland; Bs, backswamp; F, alluvial fan; f, channel fill; La, lateral accretion; Lg, lag concentrate; N, natural levee; O, oxbow lake; I, bank accretion; S, flood-plain splay; Va, vertical accretion. From Vanoni (1971), by permission of Am. Soc. Civil Engrs.

the bed load temporarily at rest. Their surface forms include ripples, waves, dunes, antidunes, and longitudinal or crossing bars. All of these features may move rapidly in the channel, but some of them may persist in varying stages of flow. Bars in the channel are *bed accretion;* bars along the banks are *bank accretion. Channel-lag deposits* are formed by removal of finer or lighter particles, leaving behind larger and heavier particles. Coarse gravel or boulder beds at the bottom of valley sediments are lag concentrations. *Channel-fill deposits* result from a general aggrading condition in the stream. These sediments may fill the channel nearly to bank level.

Channel-margin sediments are *lateral accretion deposits,* which form on the convex sides of bends as point bars (fig. 4.1). As the bend shifts toward the concave side, the point bar grows in the same direction. Depth is reduced, and velocity at high-stage flow decreases across the grow-

ing bar, which induces deposition of sediment, and the bar may grow bank high.

Two kinds of overbank sediments are vertical accretion deposits and flood-plain splays (fig. 4.1). *Vertical accretion deposits* form by deposition of suspended load during overbank flooding. As velociity is abruptly checked during bank spillover, the thickest and coarsest particles are deposited, forming low ridges or *natural levees* bordering the channel. Away from the channel, finer-textured sediment is deposited in the *backlands.* Where ponding occurs in *backswamps,* the finest-textured sediments build up vertically. *Flood-plain splays* are relatively coarse sediments that spill through a breach along the channel bank and form in a fanlike pattern over finer sediment of the flood plain.

Identification of channel, channel-margin, and overbank deposits is difficult unless they are specifically associated with a well-defined feature. A

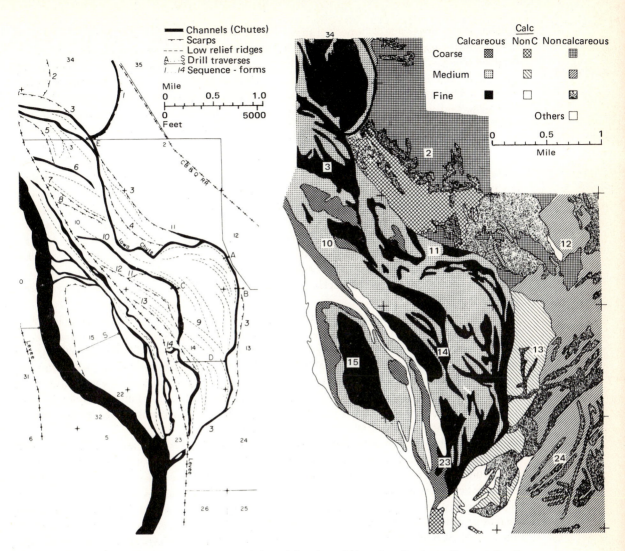

Figure 4.2 Geomorphic features and soils on a flood plain of the Missouri River, Otoe Bend area, Fremont County, Iowa and Otoe County, Nebraska. From Ruhe (1971), by permission.

point bar on the convex side of the bend of an existing channel is readily recognized, but how is it identified after the channel has migrated laterally and vertical accretion deposits cover the bar?

Recall the complex braided pattern at Otoe Bend on the Missouri River in the late 1930s (fig. 3.16). That area was covered by seven floods between 1942 and 1952. Channel accretion deposits were buried by vertical accretion sediments deposited during those floods (fig. 4.2). The area now has an elliptical saucerlike form, with a rim of coarser sediment, an intermediate band of medium-textured sediment, and a central area of finer sediment. The rim is 3.5 feet higher than the central basin and is a natural levee adjacent to the main channel and chute. The central basin, which has finer sediment, is backland or backswamp.

Note distinct textural changes in five cores across the area (fig. 4.3). At sites 4 and 8 on the natural levees, sand and silt dominate clay in the upper parts of the cores. At sites 5, 6, and 7 in the basin, silt and clay dominate sand in the upper parts of the cores. Note at these sites the reversal of texture in the lower parts of the cores, where sand and silt dominate clay. These lower sediments are channel accretion deposits (fig. 3.17), and the upper sediments are vertical accretion deposits (fig. 4.2). To bore out the original mosaic of channel deposits would be extremely difficult. Valley sediments are complex in themselves, and complex gradational relations exist between the various types (Vanoni, 1971).

Geomorphic features of flood plains

A flood plain slopes down valley, but across a valley a flood plain is not flat but is essentially horizontal. There may be local relief across the flood plain, but if elevations of topographic highs and lows are plotted against distance and a line is fitted to the points by least squares, the regression line is horizontal (Ruhe, 1967). If a significant

slope constant enters the regression equation, true flood-plain expression may be suspect.

An association of landforms comprises the bottom landscape in valleys. The Missouri River Valley from Sioux City, Iowa, to Kansas City, Missouri, has three divisions across valley (Glenn et al., 1960). Fresh meander scars, abandoned channels, and flood-plain scrolls are in a *channel belt* one to two miles wide along the present channel. The channel belt indicates the migration pattern of the 1879, 1890, and more recent channels of the Missouri River (fig. 4.4). The next outer band in the valley is the *meander belt*. This contains abandoned channels or *oxbows,* which are older than 1879 and may be considerably older. Water in the abandoned channels forms *oxbow lakes.* Beyond the meander belt toward the valley walls is the *flood basin*.

The freshness of form of a geomorphic feature relates to its age. Within the channel belt, flood-plain scrolls are mappable features, although the relief is only one to two feet (fig. 4.2). *Scrolls* are crescent-shaped landforms with coarser sediment than adjacent features. They are the most prominent soil patterns in the channel belt (fig. 4.2).

Chutes and scarps are common on flood plains. A *chute* is a narrow, curved, or sinuous channel that may or may not contain water crossing the flood plain. A *scarp* is an abrupt slope change that crosses the flood plain in a curved pattern, and land elevation is higher on one side than on the other side of the slope. The significance and origin of chutes and scarps can be determined from a sequence of maps and air photographs of an area. There are 15 maps of Otoe Bend dated from 1852 to 1970, and 11 air photographs dated from 1925 to 1966. By chronological comparison, one sees that chutes (1) form in bed and bank accretion in the stream channel, (2) form at and along the contact between bank accretion and the bank, (3) form in bank accretion as the main channel shifts away from a bank,

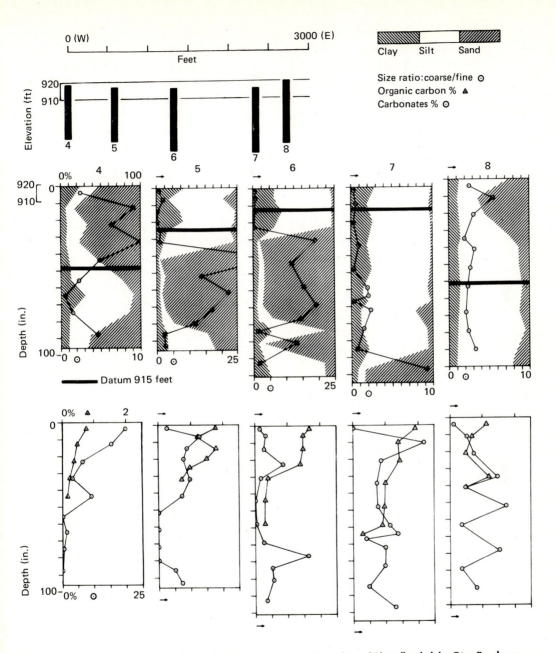

Figure 4.3 Characteristics of sediments and soils along a traverse *S* on Missouri River flood plain, Otoe Bend area.

(4) form in accretion deposits both near and farther away from the main channel, (5) are never the main channel of the stream, and (6) become landlocked by stabilization of bank accretion and concentration of the main channel away from the bank. The button-hook chute (fig. 4.2, landform 14) gradually formed after 1919 and became landlocked in 1930.

Following a similar methodology, scarps can be interpreted. What is now a scarp (landform 3) was the left bank of the Missouri River between 1879 and 1890. Its height varies from two to six feet. Another scarp (landform 13) was the left bank of the Missouri River in 1923, and the part of it that joins the chutes (landforms 10, 14), was the left bank in 1928. Southward, the left bank of the chute was the left bank of the Missouri River in 1923 and 1926.

The fit of landforms of the flood plain to historic maps and air photographs also provides a chronological framework for the soils of the flood plain.

Alluvial soils

Note the major separation of soils along the 1879 and 1890 left bank at Otoe Bend (fig. 4.2). Soils to the east in the meander belt are noncalcareous, but soils to the west in the channel belt are calcareous. Along the left bank is a narrow zone of soils where calcareous material overlies noncalcareous material. This is the spillover zone of 1879 and 1890, where natural levees and floodplain splays buried pre-existing soils of the meander belt.

Within the channel belt, soils have light-colored surface horizons, and their subsoils are stratified showing little pedogenic weathering. These soils have AC profiles. In the meander belt the soils have thicker, dark-colored surface horizons, and their subsoils have color contrasts and pedogenic structure. These soils have ABC profiles.

These contrasts illustrate a common problem in working with soils on flood plains: What prop- erties are sedimentary features or pedogenic features in the group of soils formerly called *Alluvial soils?*

On the Wolf Creek flood plain in northeastern Iowa, sediments forming low-relief ridges are coarse-textured, but intervening depressions have finer-textured sediment. Ridge soils have little profile development except for accumulated organic carbon, but depression soils have clayey B horizons with strong subangular blocky structure. Laboratory data support these observations, but lead to other problems (table 4.1). Both soils have sedimentary stratification, as is shown by the layering of sand (greater than 62μ). The ridge soil has little vertical organization of clay (less than 2μ), but the depression soil has excellent vertical arrangement. Clay content is low in the A2 horizon but increases through the B1 to a maximum in the B22 horizon and then decreases through the B3 to the C horizon. This arrangement is characteristic of a clay-pan soil or *Planosol*. Any decrease or increase in clay is balanced by opposite variations in sand and silt.

There is little weathering in the ridge soil. Base saturation (table 4.1) shows little leaching. Organic carbon has accumulated according to a normal distribution with increasing depth in the profile. Cation distribution and base saturation in the depression soil show leaching of bases in the A1 and B1 horizons, which supports the view that the clayey B horizon may be pedogenic. Organic carbon distribution is normal and not diagnostic.

To study the clay in the B horizon, a technique known as *micromorphology* is used. Thin sections are made of plastic-impregnated samples of soil horizons and are examined under the petrographic microscope (Brewer, 1964). If clay, iron oxide, or organic carbon accumulate in a soil B horizon they commonly coat the surfaces of soil aggregates and also coat the interior walls of voids (fig. 4.5). The coatings appear in cross section as layers that parallel the covered surface and are known as "clay skins," "cutans," or "argillans."

Figure 4.4 Air photograph of a flood plain of the Missouri River above Omaha, Nebraska: B, backland or flood basin; CB, channel belt; MB, meander belt; T, terrace; Tb, tributary valley; U, upland. Scale is in miles.

Note the contrast in the thin sections of the *B* horizons of the soils (fig. 4.5). The ridge soil has sand and silt grains randomly dispersed in a finer-grained matrix that is impregnated with iron oxide and organic carbon. There is no alignment or orientation of clay. The depression soil has clay skins, as is shown by the swirling pattern of light and dark material of clay, iron oxide, and organic carbon. Clay skins demonstrate pedogenic processes of eluviation and illuviation (fig. 2.11). Some of the clay in the *B* horizon is pedogenic.

Most geomorphic features on flood-plains have low relief, usually with accompanying poor surface and subsurface drainage, and shallow water tables. Soils inherit the poor internal drainage, and gleying of soils is common (fig. 2.11).

Flood-plain surveys

Because of man's increasing use of flood-prone areas, flood-plain surveys are becoming more important. The information obtained regarding the depth, velocity, frequency, and areal extent of flooding of a valley floor throughout the range of possible floods is needed as a guide for zoning and planning urban and rural use of flood plains.

Methods for mapping flood-prone areas are: physiographic, pedologic, botanical, occasional

Table 4.1 Soils on Wolf Creek flood plain

Horizon	Depth	Particle size				Cations			Base Saturation	Organic carbon
		greater than 62μ	62-16μ	16-2μ	less than 2μ	H	Bases	CEC		
	(in.)	(%)	(%)	(%)	(%)	(me/100g)			(%)	(%)
Soil on low ridge										
A11	0-8	40.5	26.8	14.6	18.1	5.2	18.8	24.0	78.3	2.58
A12	8-17	52.8	24.3	9.1	13.8	4.7	13.4	18.1	74.0	1.89
A3	17-25	33.0	34.6	13.3	19.1	6.3	18.0	24.3	74.0	1.91
B1	25-35	35.7	34.1	12.4	17.8	6.3	16.4	22.7	72.2	1.60
B2	35-44	36.3	33.0	12.6	18.1	5.7	16.4	22.1	74.2	1.28
B3	44-53	29.9	29.8	14.0	18.4	5.0	16.7	21.7	77.0	1.16
C1	53-60	51.1	27.3	8.7	12.8	3.1	10.8	13.9	77.7	0.53
C2	60-66	69.8	17.2	4.0	8.9	2.0	7.0	9.0	77.8	0.37
Soil in adjacent depression										
A1	0-5	6.1	32.5	35.3	26.2	11.7	10.1	21.8	46.3	2.30
A21	8-11	14.7	33.6	35.0	16.7	10.0	8.2	18.2	45.1	1.05
A22	11-15	18.7	30.8	29.9	20.5	8.0	11.1	19.1	58.1	0.81
B1	15-18	10.6	31.6	25.7	32.2	8.3	19.8	28.1	70.5	0.70
B21	18-26	7.3	24.0	25.4	43.2	8.0	28.0	36.0	77.8	0.63
B22	26-33	5.7	16.5	29.1	48.7	7.0	35.9	42.9	83.7	0.42
B23	33-38	4.8	28.4	26.6	40.1	4.2	27.3	31.5	86.7	0.33
B31	38-45	1.6	45.4	27.2	29.6	3.3	21.8	25.1	86.9	0.22
B32	45-50	11.0	41.2	20.4	27.4	2.6	18.7	21.3	87.8	0.19
C1	50-55	50.5	25.1	8.9	14.8	1.4	9.0	10.4	86.5	0.15
C2	55-58	27.6	44.9	11.5	16.0	1.6	10.2	11.8	86.4	0.19

flood, regional flood of selected frequency, and flood-profile and backwater curves (Wolman, 1971). *Physiographic mapping* delineates topographic features of a flood plain on a map (fig. 4.2) and correlates them with flood discharges of known frequency. Specific topographic levels of the flood plain are shown which will be under water during floods of known frequency and elevation. The map is a mosaic of patterns of *A* unit for recurrence interval *a* and elevation *x*, of *B* unit for recurrence interval *b* and elevation *y*, and so on.

A *soil survey* of flood plains delineates different kinds of soils on a map. An accompanying legend provides estimates of soil permeability and surface and subsurface drainage conditions of each soil. The soil map presents a picture of various grades of wetness of the flood plain but does not depict coverage by water during floods of known frequency and elevation. As in physiographic mapping, correlation is made between soil patterns and flood discharges. Topographic and soil maps are complementary, but hydrologic relations must be established before these maps are valuable as tools in a flood survey.

Belts or *zones of vegetation* may indicate flood levels on a flood plain, but usually the relation is not definite, and consequently the information cannot be relied upon in detail. "Flood-training" of cottonwood saplings corresponds only roughly to median height of annual peak discharge along the Little Missouri River in North Dakota (Everitt, 1968). Vegetation patterns may be more important in reconstructing the historic formation of the flood plain than in depicting flood frequencies and magnitudes.

Occasional flood is based on the establishment of flood lines seen in air photographs, marks of historic floods on the ground and on structures, and generalization of flood heights from stream-gauging data at a number of stations. Occasional flood does not have a specified frequency or recurrence interval. In the United States, maps of "approximate area occasionally flooded" are

Figure 4.5 Photomicrographs of soil thin sections. Scale: 2 in. = 1 mm. From Ruhe (1969a), by permission. © Iowa State University Press, Ames.

generally on seven-and-a-half-minute topographic sheets at a scale of 1:24,000 (see Map of Flood-Prone Area, Owensboro West Quadrangle, Kentucky-Indiana, U.S. Geological Survey).

Regional flood is based on measurements at stations of the heights above the channel bed attained by floods of different magnitudes. Flood heights for a given flood magnitude are then mapped. An "intermediate regional flood" along Clifty Creek near Columbus, Indiana, covers the flood plain from valley wall to valley wall (fig. 4.6). Above mile 3, this regional flood is slightly higher than an actual flood during January, 1959. A "reasonable maximum flood" (100-year flood) simply rises up the valley walls (fig. 4.7). In calculating a regional flood in a drainage basin,

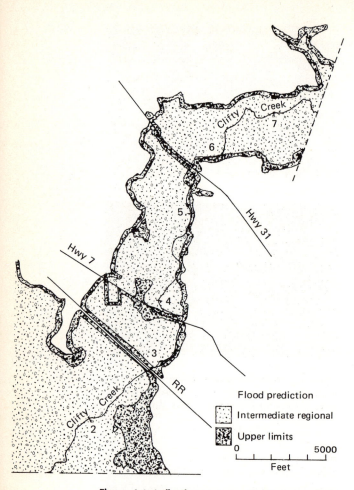

Flood prediction

Intermediate regional

Upper limits

0   5000

Feet

Figure 4.6 A flood-survey map for the area along Clifty Creek near Columbus, Indiana. From Corps of Engineers (1968).

slope increases in each stream above the junction and then flattens. The effects of man-made structures must also be analyzed as part of backwater influence, since structures on the flood plain affect flood height and rate of flow. Backwater and other effects are related to the flood height, which permits estimate of the flood line at specific magnitudes and frequencies.

Flood-plain surveys are applied geomorphology. In the United States in 1971, about 4000 urban communities required information on flooding and extend of areas subject to flooding (Wolman, 1971). The need for such information is great, and it will increase as flood plains are subjected to greater use.

Terraces

A *terrace*, a feature contained in a valley, is bordered on one side by a higher recognizable valley wall and on the other side by a scarp, that descends to some lower level. A terrace may be bounded on the valley-wall side by a higher terrace so that a sequence of stepped surfaces form from the flood plain to the valley wall. Each terrace is like a tread of a staircase, and each intervening scarp is like a riser. A terrace is usually a surface form of unconsolidated material carved from valley fill (Tator, 1953). The tread of the terrace across the valley is similar to the transverse profile of the flood plain and is essentially horizontal but not necessarily flat. Terraces may have relict flood-plain features such as abandoned channels, scrolls, levees, and backswamps.

Many terms are applied to terraces (Leopold and Miller, 1954; Howard, 1959). If terrace levels match from one side of the valley to the other, the terraces are *paired*, otherwise they are *unpaired*. A *cut terrace* is cut into and below higher alluvial-fill level. A *fill terrace* is similar except that after the cutting, filling occurs but to a lower level than the original alluvial-fill level. If a valley-confined cut surface is on bedrock, it is a

flood heights of different recurrence intervals are related to the drainage area, and curves are prepared that estimate flood heights at ungauged sites of the known drainage area.

Backwater influences in flood-plain mapping are the effects of one stream on another. Where the Racoon River joins the Des Moines River in Iowa, flood profiles of both streams show the effect of backwater from one stream on the other at various times during flooding (Myers, 1963). The bench (Tator, 1953) or a *strath terrace* (Howard, 1959). Terrace nomenclature may be visualized diagrammatically (fig. 4.8).

Problems in mapping and correlation

The true flood plain must be separated from the lowest terrace in a valley. If a terrace represents a former valley floor, that level must once have been flooded. One should not consider as a terrace a level that has been flooded during the present stream regimen; any level within the valley covered by flood water of the present stream should be excluded. The levels subject to flooding are called flood-plain steps (fig. 4.8) (Howard, 1959). What frequency and magnitude of flood disqualifies a terrace (Wolman and Leopold, 1957)? Any surface frequently modified by scour

and/or deposition should not be termed a terrace.

Using the rule that the surface of a terrace across a valley need not be flat but must be essentially horizontal avoids many problems in terrace identification. If a surface continuously rises toward a valley wall, it may be an alluvial fan or a valley-flanking pediment (Frye and Leonard, 1954). Along the Rio Grande in southern New Mexico, a valley-flanking surface often is called the Picacho Terrace, but it rises continuously away from the valley axis. If elevations (Y) are plotted against distance (X) and a curve is fitted by least squares, the regression line at some places is expressed as $Y = a + bX$, at other places by log $Y = a + bX$ (Ruhe, 1967). The Picacho surface is not a terrace but a valley-flanking alluvial fan and pediment complex.

A terrace may be modified by erosion or dissec-

Figure 4.7 Flood profiles along Clifty Creek near Columbus, Indiana. From Corps of Engineers (1968).

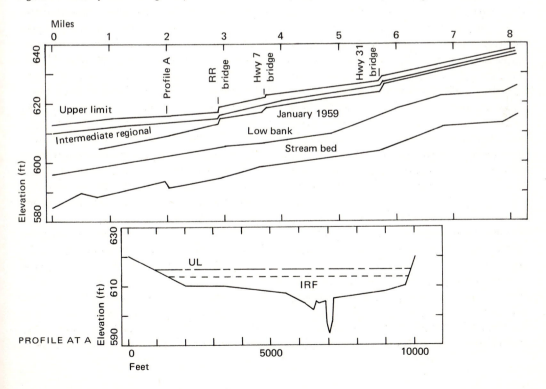

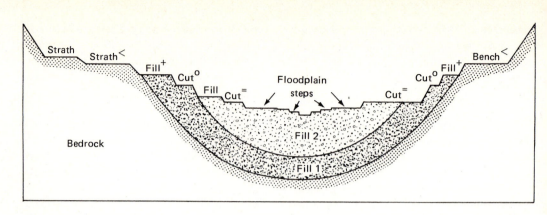

Figure 4.8 Nomenclature of terraces. From Howard (1959), by permission. © University of Chicago Press, Chicago.

tion. Original remnants may only be preserved along summits between drainage lines. To reconstruct a terrace, remnants are put together like a patchwork quilt in a morphometric analysis.

If the original terraces remain along the valley walls, a longitudinal profile can be reconstructed down valley. The profile may parallel or diverge or converge with profiles of other terraces, the flood plain, and the stream channel. Consider the Boyer River Valley in western Iowa. About 23 miles above the junction of the river with the Missouri River, and farther upstream, a terrace stands above the flood plain (Daniels and Jordan, 1966). Below mile 23, the terrace is buried by flood-plain alluvium. When the profiles are fitted by least squares, the profiles cross; they show that the lower part of the Boyer Valley previously was not filled to the present flood-plain level and that the Missouri River Valley at the mouth of the Boyer was not either (Ruhe, 1969a). A comparison of the calculated profiles showed that the lower Boyer Valley filled about 33 feet above the terrace gradient.

A fill terrace has an associated sediment body. The surface and body are traced across and up and down the valley, using morphostratigraphic methods. A cut terrace, on the other hand, is an erosion surface that can cross different materials

(fig. 4.8), and it can be traced only morphometrically.

Significance

Terraces demonstrate episodic formation of valley landscapes. They mark previous levels of valley fill and stream planation. Stream regimens changed, and the causes of these changes are involved in the interpretation of terraces.

Tectonic movements such as uplift, tilting, faulting, or warping regionally or locally cause an increase in stream gradient. Velocity increases and adjustments occur in the flow system of $Q = wdV$. In Willow River (fig. 3.12), an increase in slope caused an increase in width and depth of channel incision. If *lateral planation* during stream meandering follows channel incision, a surface is cut below the original level, and a terrace forms.

Eustatic change is a worldwide rise or fall in sea level. During the Pleistocene epoch, glacier ice accumulated on the continents, and sea level was lowered since the water bound in ice did not return to the sea. During the last glacial maximum, sea level was about 350 feet lower than it is now (Donn et al., 1962). Streams cut down to that level. With deglaciation, melt water returned to the sea. Sea level rose, which caused valley filling. The formation of terraces in the lower Mississippi River Valley are thus explained (Fisk, 1944). Five cutting episodes were related to repeated glaciations. Four fill terraces, Williana (the highest), Bentley, Montgomery, and Prairie (the lowest), were related to intervening deglaciations. The flood plain relates to a fifth, the present deglaciation (McFarlan, 1961).

There is a controversy regarding eustatic change at a stream's mouth and its impact upstream. If sea level lowered and raised several hundred feet, how far up valley would incision and fill occur? The terrace sequence is traced by "lower valley" people almost 700 miles to Illinois. There they are opposed by the "upper valley" side (Leighton and Willman, 1950), who claim "the lower valley" side uses incorrect correlation, incorrectly drawn structural contours, and incorrect planes of reference for purposes of interpretation. On one side of the argument, the up-valley impact is considered great, but on the other side, it is not. It is difficult to conceive that the lowering of the river mouth 350 feet at the Gulf should cause a procession of knickpoints hundreds of miles up the Mississippi, Missouri, and Ohio Rivers. Adjustments in this viewpoint are called for in near coastal areas.

Climatic change may trigger a change in the stream regimen. An increase in rainfall causes an increase in discharge, which requires adjustments in width and depth. A decrease in rainfall causes a decrease in density or kind of vegetative cover, which in turn causes an increase in discharge. Capture of new drainage area by headward extension of a stream adds discharge to a system without any climatic change. During glacial times, ice crossed drainage divides, and melt water added discharge to watersheds.

If a terrace can be formed in many ways, how is the origin of the feature interpreted? The problem can be resolved only by determining the physiographic, structural, and environmental history of a region and fitting the terrace system within that framework.

The age of terraces decreases from the highest to the lowest level. Each terrace in a sequence has experienced all of the natural history of all the terraces below it. Recalling the soil function $s = f(cl, o, r, p, t . . .)$, one may expect a sequential relation of soil development on terraces (Walker and Hawkins, 1958). A sequence of terrace soils is an excellent model for systems analysis within the physiographic and environmental framework.

Alluvial Fans

An *alluvial fan,* a landform composed of alluvium, is shaped somewhat like part of the surface of a geometric cone. Alluvial fans may occur anywhere that a stream descends to a plain; they are common along valley slopes in subhumid Iowa (Daniels and Jordan, 1966) and the humid Shenandoah Valley in Virginia (Hack, 1965), but they are better known as landforms of semiarid and arid regions (Denny, 1965), although they are not exclusive features of "arid" or "semiarid" geomorphology (Bull, 1963).

Form and composition

The external form of an alluvial fan may be visualized geometrically as a cone that is generated by a line moving about a fixed point on that line. A cone has an external linear gradient from the apex to the base, but an alluvial fan generally has a curvilinear gradient. The form may be described by partially revolving a curve about a line generating a surface of revolution (Troeh, 1965).

Planimetrically, the contours of an alluvial fan are concentric arcs, and some fans have excellent symmetry. (See Cedar Creek Alluvial Fan on the topographic map, Ennis Quadrangle, Montana, 15-minute series, U.S. Geological Survey, 1949.)

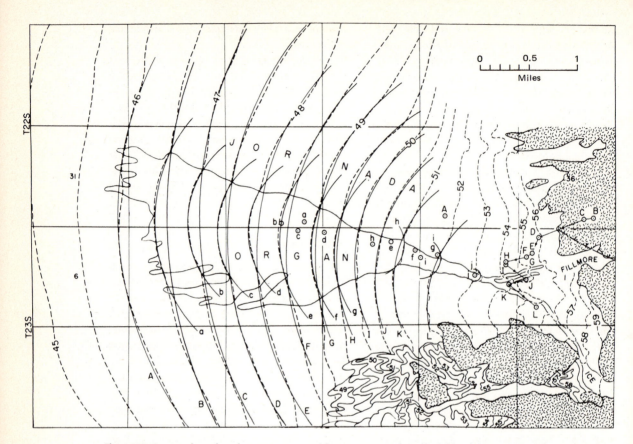

Figure 4.9 Inset and overlay of a younger on an older component of an alluvial fan using constructed arcs. Contours in hundreds of feet. From Ruhe (1967), by permission of N. Mex. Bur. Mines and Miner. Resour.

Contours represent curves formed by the intersection of a family of parallel horizontal planes with the land surface. A method of morphometric analysis of an alluvial fan is fitting arcs of curvature to contour lines (Ruhe, 1967). Centers of radii of curvature are located by constructing perpendiculars to tangents to the arc, and in turn locating a center of symmetry for each contour. Centers displace on natural alluvial fans, and arcs of all contours cannot be circumscribed from a single point. The shifting of centers indicates that the loci of dispersion of sediments that build the fan also shift.

This curve-fitting technique is also useful in de-

termining the inset and overlay of a younger on an older fan (fig. 4.9). Two geometric sets are needed. One has larger radii and lesser curvature, marking the older surface. The other has smaller radii and greater curvature, marking the younger fan. The inset of the younger fan below the older one is shown by reversed curvature, and the overlay of the younger fan on the older one is shown by greater curvature of contours of the younger fan. Note these relations between two fans by comparing the two contour sets from 4650 to 4950 feet.

Where curvature reverses, an upslope erosion surface is cut below the older fan. Downslope a constructional younger fan is built on the older fan. Soils on the surface of the older fan pass beneath the younger-fan sediments as *buried soils*, or *paleosols*, which are common features in alluvial fans in the southwestern United States (Ruhe, 1965a).

The radii of an alluvial fan must cross contour lines at right angles, and on a map they curve from the apex downslope to the fan margin. Drainage lines on a fan surface generally approximate the positions of the radii (fig. 4.10).

The profiles along the radii are similar to the concave upward longitudinal profile of a stream. A profile down a radius from Fillmore Canyon in southern New Mexico (fig. 4.10) has progressively shallower gradients of 1308, 667, 400, 355, 157, 127, 125, 77, and 71 feet per mile (Ruhe, 1964a). If curve fitting is used where X is distance and Y is elevation, the profile above 4800 feet is described by the hyperbola of the form $Y = 1/(a + bX)$, below 4800 feet by an exponential curve of the form $\log Y = a + bX$ (Ruhe, 1967).

Radii, curvature, and profiles may indicate multilevels on a fan as shown by inset and overlay and may aid in explaining sediment distribution. Note the kinds of rocks in the source area of the Fillmore Canyon fans (fig. 4.10). Also note the composition isopleths that show lateral changes associated with diverging radii of the fans. Rhyolite gravel decreases to the north, and monzonite gravel decreases to the south. Andesite gravel is dominant along the axis of the fan.

The area of an alluvial fan (Af) is related to the size of the source area above the fan apex (Am), as expressed by a power function $Af = kAm^e$ (Denny, 1965). Gradients of alluvial fans also relate to the size of the drainage area as a power function or to the basin area and the relative relief of the drainage basin (M. A. Melton, 1965). Relative relief is expressed in a ruggedness number of $H/\sqrt{A}$ where H is the vertical relief above the fan apex and A is the drainage-basin area.

Alluvial fans are composed of various kinds of sediment ranging in particle size from clay to boulders. In loess country in western Iowa, fans along the base of valley walls are composed mainly of silts and do not differ texturally from the source loess of the uplands. In the western and southwestern United States, where size and relief are much greater, the particle-size range is also greater. Sediments vary from mudflows to water-laid sands and gravels (Bull, 1963) to debris flows (Beaty, 1963) that contain boulders larger than a 4 x 4 jeep (Ruhe, 1967).

Size sorting of particles occurs down gradient in alluvial-fan sediments. As distance from the sediment source increases, the slope gradient progressively decreases, and transportability adjusts to changing slope. On fans in Death Valley, California, and in southern Nevada, mean particle size (Y) decreases with distance (X) as expressed by the equation $\log Y = a - bX$ (Denny, 1965; Bluck, 1964). In southern New Mexico size also decreases with distance downslope from a source, but the equations differ, being $Y = a - bX$ and $Y = a - b \log X$ (Ruhe, 1964a, 1967).

Origin

The basic requirements for the formation of an alluvial fan are a standing land mass and a stream descending to a lower surface of flatter gradient. At the gradient change, the transported load is dumped, and it spreads outward in a fanlike pattern. Deposition backs up the gradient, building the curved partial cone. A stream may descend any radius and continue to build the fan to the left, the center, or the right. Building is oriented along washes or stream courses (Denny, 1965) and may involve *debris flow,* which is the transport of coarse particles in a viscous mass of finer particles and water (Beaty, 1963). Debris flow is probably the only mechanism that accounts for large boulders in fan sediments many miles from the rock source (Ruhe, 1967).

Most large alluvial fans in the southwestern United States are not single landforms built as one

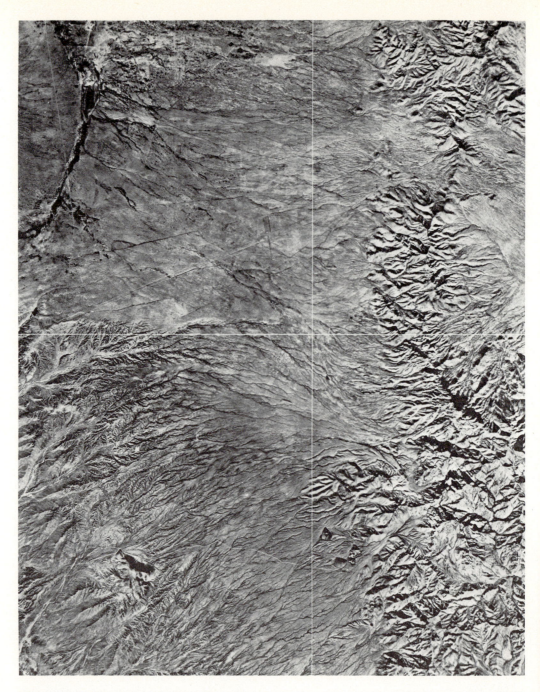

Figure 4.10 Air photograph of alluvial fans from Organ Mountains near Las Cruces, New Mexico, showing sediment dispersion from source-rock areas of mountains. Numbers 5 through 90 indicate the percentage of gravel of given rock species. From Ruhe (1964a), by permission of Assoc. Am. Geog.

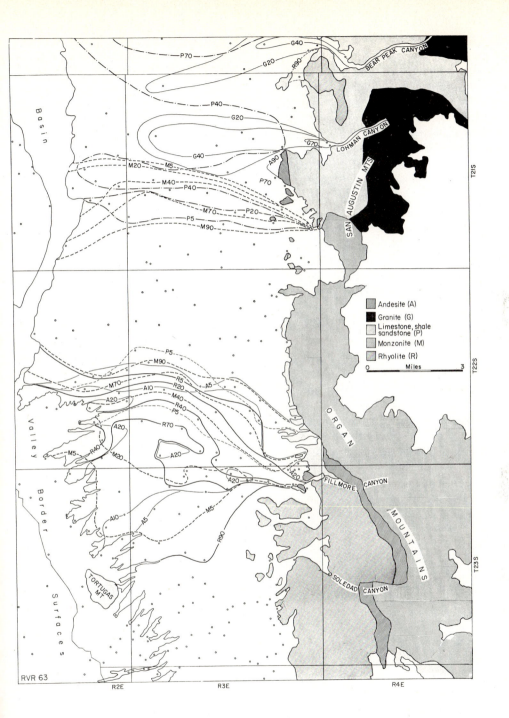

RVR 63

deposit but consist of both older and younger parts. The Fillmore Canyon fan in southern New Mexico has older and younger parts (Ruhe, 1964a), and different kinds of soils are on them (Ruhe, 1967). Younger parts have weakly developed soils, while older parts have strongly developed soils with red, textural B horizons and indurated carbonate horizons (fig. 2.13). In Arizona "red-soil fans of Frye Mesa type" have red soils with clayey B horizons that stratigraphically separate deposits of different ages and types. The red soil is on the surface in places but is buried beneath younger deposits in other places (M. A. Melton, 1965). In the desert regions of California, alluvial fans have "oldest, older, and youngest fan surfaces" (Hooke, 1967). These alluvial fans were constructed during episodes of varying dates.

Alluvial fans in Death Valley, California, apparently are similar. Among three major mapping units on Shadow Mountain fan, map patterns truncate one another, and air photographs show that one pattern is inset below another. Similar relations occur on Hanaupah Canyon fan, where the three major mapping units are described as weathered gravel commonly cemented by caliche, varnished gravel, and unweathered gravel (Denny, 1965). Presumably, three different kinds of soils relate to the three different parts of the fan.

Is it feasible then, that the origin of alluvial fans in Death Valley "is related to some sort of balance or interaction between mountain degradation and valley alluviation—perhaps to a condition of dynamic equilibrium between a fan building wash and its surroundings—rather than to the fan's stage of development in some evolutionary sequence" (Denny, 1965, 1967)? Can the evolutionary sequence be avoided?

If alluvial fans are in dynamic equilibrium (Lustig, 1965), a balance must exist between the erosion in the source area and the deposition of the fan sediments, and the rates of erosion and deposition must be equal. The correlation of the size of the source area and the size of the fan is not direct

evidence of a balance of processes. Relict features of some parts of alluvial fans show lack of equilibrium with modern processes. Application of dynamic equilibrium requires proof that a steady state has been attained. One cannot assume that the condition exists and that the present processes are identical to those of the past. Application of dynamic equilibrium in New Mexico has major limitations when long time spans of fan origin are involved (Hawley and Kottlowski, 1969). Extensive preservation of relict landscapes and soils and paleoecological evidence indicate that cyclic shifts in climate affected landscape evolution.

What causes different episodes of alluvial-fan building? It is not enough to state that a fan "is in the Davisian stage of youth" because it is growing (Beaty, 1970). The triggering of erosion and sedimentation of alluvial fans is like that of stream terraces; that is, they are triggered by tectonic uplift (Bull, 1964) or climatic change (Lustig, 1965; M. A. Melton, 1965; Hawley and Kottlowski, 1969).

Up to this point we have focused attention on a stream, a channel, or a valley. In nature these things are grouped or arranged in an organized system. An interconnecting system of drainageways is a *drainage net,* which consists of a main channel and its tributaries. The planimetric design of the net is the *drainage pattern*. A *drainage basin*, or *watershed*, is a part of the earth's surface that is drained by a main stream and its tributaries and that has a divide separating it from another basin. All surface waters discharge from the drainage basin via the drainage net and the mouth of the main stream. All sediment derived by erosion within the drainage basin and not held within it is carried through the mouth. The drainage basin is a fundamental geomorphic unit of the land, and all flow of surface water is governed by its properties.

Analysis of drainage nets

Drainage nets have dendritic, parallel, rectangular, radial, and centripetal patterns, and the patterns often are combinations (Howard, 1967). The dendritic drainage pattern is like that of limbs branching from the trunk of a tree. In the parallel pattern, main streams have parallel courses, as do tributaries. In the rectangular pattern, stream bends and junctions are approximately at right angles. In the radial pattern, streams extend outward along radii from a common center like the ribs of a fan. The centripetal pattern has streams that descend from a periphery to a common central depression. These patterns are of geomorphic significance (table 5.1).

5

Drainage Nets and Basins

Table 5.1 Significance of basic drainage patterns

Pattern	Significance
Dendritic	Horizontal, uniformly erodible sediments or uniformly resistant crystalline rocks with gentle regional slope
Parallel	Usually on steeper slopes but also between parallel landforms such as moraine, loess, or sand ridges
Rectangular	Joints and faults at right angles or angularly arranged landforms (usually denotes structural control)
Radial	Volcanoes, domes, erosion remnants, alluvial fans, deltas
Centripetal	Craters, playas, pond or lake basins, or other depressions

From Howard (1967), by permission of Amer. Assoc. Petrol. Geol.

Gross patterns

Drainage nets are analyzed by their gross patterns or by their components. Gross patterns may show structural control, as does an elongate, radial drainage pattern on the Dona Ana Mountains in southern New Mexico (Ruhe, 1967). The net descending to the Rio Grande trench to the west is denser than that descending to the southern Jornada Basin to the east (fig. 5.1). The curved axis of the basin parallels the smoothed contour lines around the mountains. Basin curvature and contours above 4350 feet are ellipsoid. An ellipse fits the 4400-foot contour with a major axis (*A-A* along *X*) oriented N40°W, as expressed by the equation $X^2/20.7 + Y^2/6.5 = 1$ and eccentricity $e = 0.828$. The fit is good statistically, with an estimate of error of 0.2 mile and a coefficient of correlation of 0.96 at the 1 percent level of significance.

A second ellipse fits the arcuate trend of the bounding basins through 198° of arc, as expressed by the equation $X^2/30.5 + Y^2/13.5 = 1$ and eccentricity $e = 0.748$. This fit is also good statistically, with an estimate of error of 0.3 mile and a coefficient of correlation of 0.95 at the 1 percent level of significance. Thus the topographies of the Dona Ana Mountains and the bounding basins are parts of the same geometric system with a

common center of symmetry marked by intersection of major and minor axes of ellipses.

Examine the orientation of the drainage lines within the quadrants of the ellipsoidal dome (fig. 5.1). Directions of stream lengths per 10° of arc are measured in each quadrant and are calculated as a frequency distribution: $\sum$ stream lengths per 10° arc/$\sum$ stream lengths per quadrant X 100 = % of streams properly directionally oriented in each quadrant. These values for the quadrants are: NE, 85 percent; SE, 85 percent; SW, 83 percent; and NW, 74 percent. The drainage pattern is properly oriented, which indicates that the Dona Ana Mountains are an ellipsoidal, structural dome (Ruhe, 1967).

The analysis of gross drainage patterns may also indicate varying ages of land surfaces. A study in northwestern Iowa that was done several decades ago (Ruhe, 1952) has had considerable textbook exposure since then (Thornbury, 1954, 1969; Leopold et al., 1964; Ruhe, 1969a). The Kansan, Iowan, and Tazewell surfaces have well-integrated dendritic drainage nets, but the Cary surface has a poorly integrated pattern (fig. 5.2). There is a pattern discontinuity between the younger Cary and the older Tazewell surfaces. This difference in surface drainage is a mapping criterion in Iowa that separates about 12,500 square miles of Des Moines drift lobe from adja-

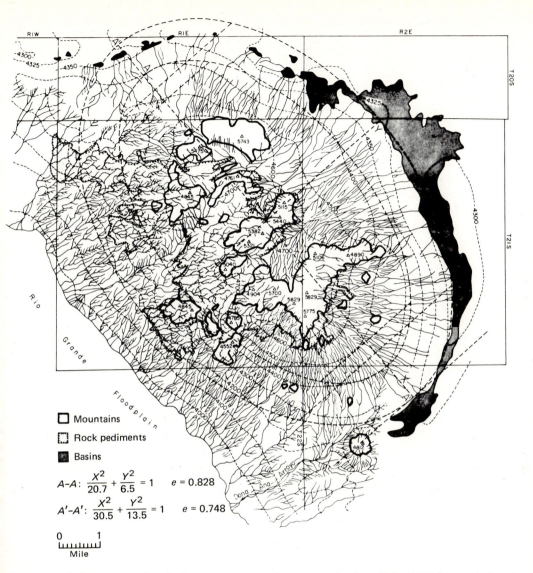

Mountains

Rock pediments

Basins

$$A\text{-}A: \frac{X^2}{20.7} + \frac{Y^2}{6.5} = 1 \quad e = 0.828$$

$$A'\text{-}A': \frac{X^2}{30.5} + \frac{Y^2}{13.5} = 1 \quad e = 0.748$$

0   1
Mile

Figure 5.1 A drainage net on a dome in the Dona Ana Mountains, New Mexico. From Ruhe (1967), by permission of N. Mex. Bur. Mines and Miner. Resour.

cent land surfaces. A similar separation occurs in Minnesota as far north as the latitude of Minneapolis–St. Paul.

Gross drainage patterns can thus be used to distinguish and interpret specific features of the earth's surface.

Components of the drainage net

A drainage net may be separated into components that permit a quantitative attack on how drainage nets and drainage basins develop (Horton, 1945). Following Horton's pioneering work, techniques and methods were improved mainly by Strahler (1950, 1952, 1956, 1958) and his associates (Schumm, 1956a; Melton, 1957; Coates, 1958; and Morisawa, 1959). Since 1950 there has been a topological explosion in watershed geomorphology, and some of its fall-out is useful.

Horton presented a system of *stream orders* in which unbranched fingertip tributaries are order 1. These join to form order 2, which in turn join to form order 3, although order 1 may also join order 3, and so on. The main stream has the highest order number. In the Horton system, selection is subjective in tracing the main stream and some lower-order tributaries to the perimeter of the watershed (fig. 5.3). Note how orders 4, 3, and 2 are traced through the drainage basin. Strahler modified the system to rid it of subjectivity (fig. 5.3). Fingertip tributaries are order 1; they join to form order 2, which in turn join to form order 3, and so on. No higher-order stream is carried through to the watershed perimeter. In the illustration (fig. 5.3), the main stream is order 4 by both Horton and Strahler numbering, and the watershed is a fourth-order drainage basin.

After numbering is completed, each order within the watershed is summed. Note a difference between the Horton and Strahler systems. There are 18 first-order, 5 second-order, 1 third-order, and 1 fourth-order streams if Horton's method is used, but there are 24 first-order, 7 second-order, 2 third-order, and 1 fourth-order streams if the Strahler system is used. Conse-

Figure 5.2 **A drainage net on land surfaces of glacial drift country in northwestern Iowa.** From Ruhe (1952), by permission of Am. J. Sci.

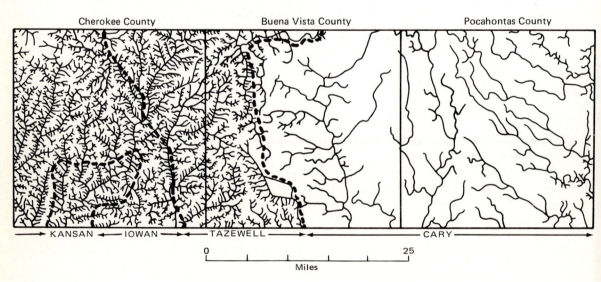

quently, *stream frequencies (Fs)* differ in the Horton relation $Fs = N/A$ where N is the total number of streams per unit area A. If the area is 2.25 square miles, the Horton Fs is $25/2.25 = 11.1$ and the Strahler Fs is $34/2.25 = 15.1$. Frequencies of individual orders also differ, as in first-order streams: $18/2.25 = 8.0$ versus $24/2.25 = 10.7$.

The reverse of stream joining is bifurcation. For example, one order 2 stream splits to form two order 1 streams. Horton's *bifurcation ratio* is the ratio of the number of streams of a given order to the number of streams of the next lower order. Using Horton's method, bifurcation ratios *(Rb)* are 3.6 for orders 2 and 1 and 5.0 for orders 3 and 2. The ratios differ in the Strahler system.

If the number of streams (Y in log scale) is plotted against the respective stream-order number (X in arithmetic scale), a straight line fits the data. The exponential curve is expressed as $Y = ab^x$ or, when fitted by least squares, as $\log Y = a - bX$. This orderliness was proclaimed by Horton as a *law of stream numbers:* the number of streams of each order form an inverse geometric relation with the order number.

Stream length is not only important in watershed hydrology as relating to drainage area (Hack, 1957), but in the Horton analysis, mean lengths of stream orders characterize the sizes of the components that comprise the drainage net. The mean length of a given order is greater than that of the next lower order and less than that of the next higher order. If stream lengths (in log scale) are plotted against their respective stream-order numbers (in arithmetic scale), a straight line fits the data. Horton therefore proclaimed a second law, which is called the *law of stream lengths:* the average lengths of streams of different orders approximate a direct geometric relation with the order number. The two laws demonstrate that a reasonable adjustment exists between the drainage net and the drainage basin.

The total length of streams in a drainage net determines its *drainage density (Dd)* in the relation $Dd = \sum L/A$ where L is stream length and A is the

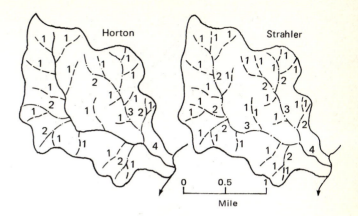

Figure 5.3 The Horton and Strahler systems of stream orders.

area of the drainage basin. The drainage density quantifies the texture of the drainage pattern. In the model drainage net (fig. 5.3), the sum of stream lengths is 11.53 miles and the basin area is 2.25 square miles. The drainage density is therefore 11.53/2.25, or 5.12, and that value can be used for comparison with other nets and basins.

Re-examine the drainage patterns in northwestern Iowa (fig. 5.2). The difference between the Cary surface and the others is obvious, but are there differences between the others? If a point-intercept method is used to determine the drainage density (Ruhe, 1969a), the values for two equal-size areas of the Kansan are 10.2 and 7.9; of the Iowan, 7.7 and 6.1; of the Tazewell, 5.4 and 4.7; and of the Cary, 2.1 and 1.9. Point-intercept or line-intercept techniques (Carlston and Langbein, 1960; McCoy, 1971) permit rapid measurement of drainage density.

The job is simplified by using an electric grid counter. An area grid with copper wires imbedded in a plastic sheet is laid over a map of a drainage net, which is then traced on the plastic sheet. The tracing pencil is wired into the electric system. At each intersection of a drainage line and a copper wire an electric contact is made, and the point intercept is recorded cumulatively by an

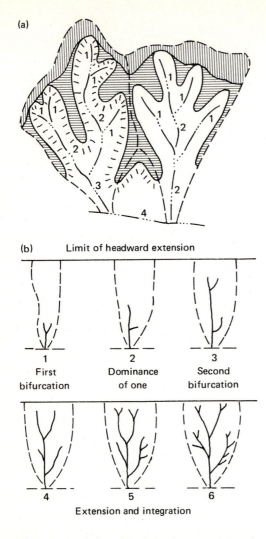

(a)

(b)   Limit of headward extension

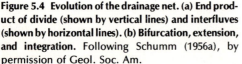

1   2   3
First   Dominance   Second
bifurcation   of one   bifurcation

4   5   6

Extension and integration

Figure 5.4 Evolution of the drainage net. (a) End product of divide (shown by vertical lines) and interfluves (shown by horizontal lines). (b) Bifurcation, extension, and integration. Following Schumm (1956a), by permission of Geol. Soc. Am.

electric counter. The area of the drainage net is also determined by the fit of squares of the grid. The drainage density, then, is $Dd = P/A$ where P is total point intercepts and A is area.

The drainage-density numbers quantify different patterns, but what do they mean? In northwestern Iowa they demonstrate progressively more complete integration of the drainage net on older surfaces.

Evolution of the net

Surface runoff follows downslope flow paths from a drainage divide to lower-level faint swales on the original land surface (Schumm, 1956a). A permanent channel forms as runoff is concentrated from an entire watershed. The most vigorously developing channel dominates its less effective neighbors and establishes itself as the axis of a drainage basin. Dominance is favored by a position along the maximum length of flow that is aligned with the watershed axis (Horton, 1945), by contact with more easily erodible material, and by structural control. This first stage of stream and valley development consists of a single central stream with a V-shaped valley (Horton, 1945). The watershed is divided along an axis, and the lateral slopes descend to the central stream (Horton, 1945; Schumm, 1956a). Tributaries form on the lateral slopes in a second stage of development. In a third stage of net development, within each half of the watershed lateral slopes descend to the tributaries and lower-order streams. A fourth stage and subsequent stages continue to evolve. The drainage pattern forms through bifurcation, channel dominance, and headward growth (Schumm, 1956a) or by extension and integration (Glock, 1931). Where two neighboring nets evolve, an upland divide stands above the limit of headward extension (fig. 5.4). Between the streams of a net and between adjacent nets, interfluves form and stand above the drainageways. Hillslopes ascend to the interfluve and divide summits. All the basic elements of the drainage basin—divide, interfluves, hillslopes, and drainage lines—have evolved.

Analysis of drainage basins

Drainage basins have geometric properties to which hydrologic, erosion, and sedimentation data may be related in a natural systems analysis. In applied geomorphology, watersheds are increasingly important as the basic unit for resource planning.

Basin characterization

Dimensional analysis is a method for studying drainage basins (Strahler, 1958), and the common dimensions, used singularly or in combination, are length, width, height (relief), area, and volume.

In *area-altitude analysis,* or *hypsometric analysis,* three-dimensional forms are reduced to two dimensions (Langbein et al., 1947). In a watershed a relative height ratio is h/H where h is the height of a selected contour above the basin mouth and H is the difference in elevation between the mouth and the highest point on the watershed divide. A relative area ratio is a/A where a is the area of the watershed above the selected contour and A is the total basin area. Measures of h/H as Y are plotted against paired measures of a/A as X. A hypsometric curve, drawn through the points, characterizes a specific basin. The *hypsometric integral* is the ratio of the area beneath the curve, from the curve to the X axis, to the total area of the plot, or the product of Y and X. Hypsometric integrals of basins may be compared.

Area-altitude analysis has many other applications. Hydrologic snow surveys generally show that with increased altitude there is an increase in depth of cover and water equivalent. Variation in annual precipitation and runoff may relate to altitude—temperature changes with altitude. The altitude of the basin above its outlet or gauging station represents a potential head of a uniform depth of water over the basin relative to the gauging station and is a factor in the rate at which water is collected and discharged. Area-altitude analysis can provide estimates of these hydrologic phenomena (Langbein et al., 1947).

The basin shape or planimetric form in part controls the stream discharge from a watershed, and simple functions express the shape. The *form ratio* relates the basin area to the square of the basin length, or $Rf = Ab/Lb^2$. The *circularity ratio* is the ratio of basin area to the area of a circle with the same perimeter as the basin, or $Rc = Ab/Ac$. The *elongation ratio* is the ratio of the diameter of a circle equal in area to the basin to the maximum basin length, or $Re = Ac/Lb$ (Schumm, 1956a).

The vertical dimension of a basin whose relief is H may also be expressed as simple functions. The *relief ratio,* expressed as $Rh = Hm/Lb$, is the ratio of the difference in elevation between the basin mouth and the highest point on the basin perimeter to the basin length. The relief ratio relates directly to stream gradients, drainage density, slope angles, basin shape, and sediment loss. In the arid southwestern United States, the mean annual sediment loss in acre-feet per square mile is an exponential function of the relief ratio (Schumm, 1954).

The *ruggedness number* is the product of relief H and drainage density D or HD, and it gives an estimate of slope steepness in a watershed. If D increases with H constant, the average horizontal distance between channels and divides decreases, and slope steepness increases. If H increases with D constant, relief increases between channel and divides, and slope steepness increases.

All of these relations about stream channels (Chapter 3), drainage nets, and drainage basins are considered "quantitative geomorphology" (Ghose and Pandey, 1963; Strahler, 1968), but many of them are no more than mathematical

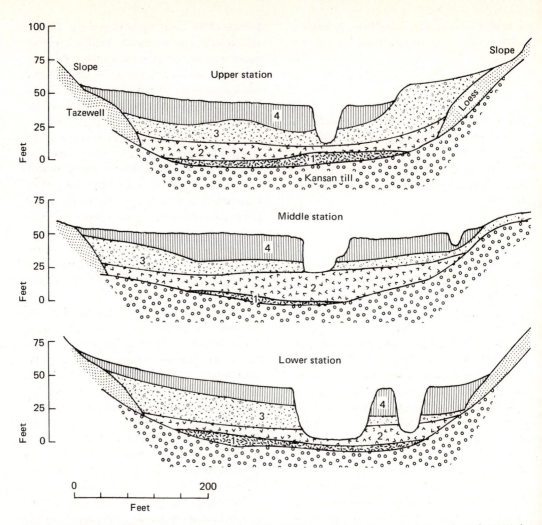

Figure 5.5 Alluvial beds in the Thompson Creek watershed, Harrison County, Iowa. From Daniels et al. (1963), by permission of Am. J. Sci.

descriptions of the external form of a part of the earth's surface. Certainly the analytical chemistry and mineralogy of a weathering profile (Chapter 2), the mechanics of landslides, and the absolute dating of land surfaces and surficial deposits cannot be excluded from quantitative work.

Basin systems

The analogy of a thermodynamic *open system* has been applied to drainage basins: matter and energy are imported and exported across boundaries, and energy is transformed uniformly to maintain a steady state (Strahler, 1952). In a watershed, rainfall is imported across the boundary. Sediment and excess surface water leave the system through the mouth of the basin. If any controlling factors change, the steady state is upset and changes occur within the system to regain the steady state. These theoretical views are invoked in *quasi-equilibrium* (Leopold and Maddock, 1953) and *dynamic equilibrium* (Hack, 1960) and general systems theory in geomorphology (Leopold and Langbein, 1962; Chorley, 1962; Scheidegger and Langbein, 1966).

Are these theoretical invocations necessary, or are they irrelevant distractions (Smalley and Vita-Finzi, 1969)? Is it perhaps better to work with natural geomorphic, hydrologic, and sedimentologic systems in watersheds, where open and closed systems can be defined within quantitative limits? In a natural *open system* a drainage basin of any magnitude is confined at its head and along its sides by the perimeter divide, but it is open at its mouth. Surface water from rainfall or melt water is collected by the drainage net in the basin and is discharged through the outlet. Sediment also is collected and concentrated along the drainage net, but only part of it is discharged from the basin through the outlet. Only a partial sediment record remains in a lower-order drainage basin (Ruhe, 1969a).

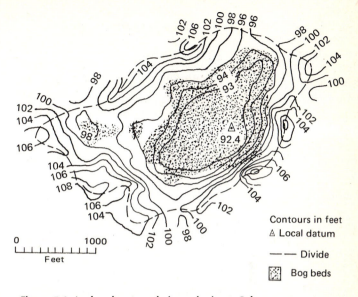

Figure 5.6 A closed-system drainage basin at Colo Bog, Story County, Iowa. From Walker (1966).

In contrast, a natural *closed system* contains an interior basin, and the drainage net descends in a centripetal pattern from the peripheral divide. Surface water from atmospheric source descends the drainageways and collects in the basin. Loss of water is only through evaporation, transpiration by plants, and subsurface percolation. Sediment is trapped in the basin, and except for what may be blown away by wind, the entire sediment record is preserved in the basin.

These two basic systems can be used anywhere (except in areas covered by perennial ice) and can be applied to any basin regardless of size. In the open system, the basin may vary from a fraction of an acre for a first-order fingertip tributary to the magnitude of the Mississippi River basin. In the closed system, the basin may vary from a fraction of an acre for a pothole to a standard basin of the Basin and Range. These two basic systems

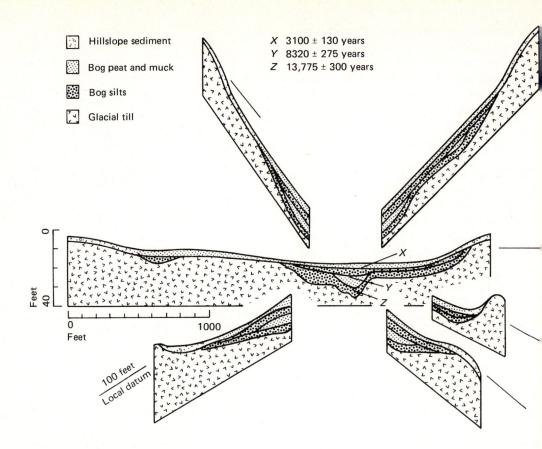

Figure 5.7 Exploded radial stratigraphic sections of Colo Bog, showing double layering. Ages were determined by radiocarbon dating. From Walker (1966).

permit study of not only processes but also earth history as contained in sediment records. The history of the earth is still a part of geomorphology.

An open-system watershed of 30 acres descends to the valley of Thompson Creek in southwestern Iowa (Daniels et al., 1963). Five alluvial beds fill Thompson Creek Valley (fig. 5.5) and contain pieces of wood or stumps of trees rooted in place, which permits radiocarbon dating. A detrital red elm log at the base of bed 3 was 2020 ± 200 years old. A willow stump in place at the contact of beds 3 and 4 was 1800 ± 200 years old. A walnut log 3 feet from the base of bed 4 was 1100 ± 170 years old. A box-elder rooted in bed 5 and 2 feet above the top of bed 4 was more than 250 years old. The oldest living tree on bed 5, dated by tree rings, is 76 years old. Putting the beds and dates together gives the following ages. Bed 3 is 1800 to 2020 years old and was formed in 220 years. The bottom quarter of bed 4 is 1100 to 1800 years old and was formed in 700 years. The top three-quarters of bed 4 is 250 to 1100 years old and was formed in 850 years. The top part of bed 5 is 76 to 250 years old and was formed in 174 years.

In the open watershed, 8 cross sections were drilled to delineate beds in the alluvial fill. The volume of bed 4 is 149,222 cubic yards, deposited between 250 and 1800 years ago at the rate of 96 cubic yards per year. The bottom quarter of bed 4 has 37,306 cubic yards, deposited between 1100 and 1800 years ago at the rate of 53 cubic yards per year. The top three-quarters of the bed has 111,917 cubic yards, deposited between 250 and 1100 years ago at the rate of 140 cubic yards per year.

Overlying bed 4 are light-colored sediments that have been deposited during agricultural use of the bounding hillslopes since the year 1850. About 7228 cubic yards were deposited between 1850 and 1965 at the rate of 63 cubic yards per year, which is the "accelerated" erosion or sediment yield caused by agricultural practices. Compare the rates. The accelerated erosion rate is less than the natural rate of bed 4 as a whole and is less than the rate of the top three-quarters of the bed. The man-induced rate is greater than the natural rate for the bottom quarter of the bed (Ruhe and Daniels, 1965). All of these rates are minimum values. The amount of sediment removed from the open drainage basin and transported to a basin of the next higher order is indeterminate.

A similar approach may be taken in a closed system (Walker, 1966). Peripheral hillslopes descend to an interior closed basin or bog where sediments derived from the hillslopes are trapped (fig. 5.6). Organic material in the sediments at various stratigraphic levels permits radiocarbon dating and establishment of a time scale. Extensive drilling and coring of the bog sediments allows three-dimensional reconstruction of the system (fig. 5.7). The volumes of the beds can be calculated.

Note that the bog sediments are paired. Peat and muck above silt overlie a similar sequence of beds. The lower silt began accumulating 14,500 years ago and stopped after 13,775 years. About 138,360 cubic yards were deposited at the rate of 191 cubic yards per year. The upper silt of 177,595 cubic yards accumulated in 5455 years at the rate of 34 cubic yards per year. The lower peat and muck accumulated at the rate of 2 cubic yards per year. These rates are true values, as the complete record is preserved.

Obviously, in this kind of systems-analysis approach (whether the systems are open or closed), where erosion, sedimentation, and their rates are involved, time-independence must be discarded. Since rates involve time and since they change, any attempt to introduce the views of steady state, quasi-equilibrium, or dynamic equilibrium are meaningless.

The sediments in the valley of the open system or in the basin of the closed system were derived from bounding hillslopes, which in turn are important components in any systems analysis of a drainage basin.

Hillslopes, which make up a large part of any land surface, are important landforms for geomorphologists, pedologists, and engineers. Hillslopes form by erosion, deposition, or a combination of both processes. When a stream cuts into a land surface, a valley is created and hillslopes form that descend to the valley bottom and ascend to the upland.

Hillslopes form by constructional processes. Glacier ice deposits till in a swell-and-swale pattern, and hillslopes intervene between topographical highs and lows. Wind deposits sand (dunes) or silt (loess) with constructional windward and leeward slopes. Water deposits sediment in an alluvial fan with an original slope. Soon after a slope is formed, it is subject to alteration by weathering, erosion, sedimentation, and mass movement. By their inherent nature, hillslopes are instable geomorphic features.

Components of hillslopes

Hillslopes may be examined both geometrically and geomorphically.

Geometric components

A hillslope is defined in space by three geometric components. The *gradient,* which is the angle of inclination of the hillslope with the horizontal plane, is always measured at right angles to contour lines, usually in degrees, by geologists, and in percent by soil scientists, or in a unit of relief per unit of planimetric length such as feet per feet.

6

Hillslopes

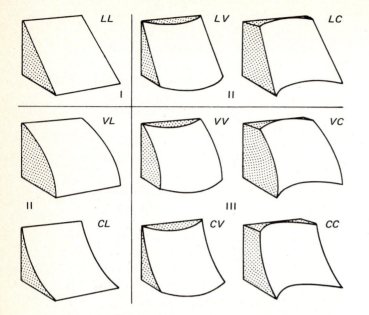

Figure 6.1 Geometric forms of hillslopes. Slope length is down the form; slope width is across the form. L, means linear; V, convex; C, concave. The simplest form (I) is colinear (LL). Group III forms, the most complex, are doubly curved. Group II forms are linear in one dimension and curved in the other.

The *slope length,* which is measured along the direction of the gradient, crosses contour lines at right angles. The *slope width,* which is measured perpendicular to the slope length, parallels contour lines.

A hillslope may be straight or curved along both length and width. If the gradient is constant per unit of length, the hillslope is straight (or linear) along the slope length. If the gradient increases or decreases per unit of length, the hillslope is curved either concavely or convexly along the slope length. If the gradient increases upslope per unit of length, the hillslope is curved concavely. If the gradient decreases upslope per unit of length, the hillslope is curved convexly. When the gradient increases or decreases downslope, the opposite is true.

The lateral shape of a hillslope is expressed by the shape of its contours and by the relations of the slope lengths to the slope width. If there is no directional change along contours, the hillslope is straight (or linear) laterally. If the contours bow outward, the lateral shape is convex; if they bow inward, the shape is concave. If there is no directional change of the slope length along slope width, the hillslope is straight (or linear) laterally, as the slope lengths are parallel and mark an inclined plane. If the slope lengths converge downward, the hillslope is concave laterally; if they diverge downward, it is convex laterally. The converse is true upslope.

These three possible shapes—linear, convex, and concave—along slope length and width yield *nine basic slope geometries* with three groups of complexity (fig. 6.1). The simplest form is straight length and straight width *(LL)*. More complex surfaces have straight length with curved width *(LV, LC)* or straight width with curved length *(VL, CL)*. The most complex surfaces have curved length and curved width *(VV, VC, CV, and CC)*.

In terms of runoff on a smooth *LL* slope, water should not channel, and sheet flow should be dominant. Recall the Manning equation (Chapter 3):

$$V = 1.49 \frac{R^{0.67} S^{0.50}}{n}$$

Uniform linear slope length should provide uniform velocity. On a smooth *VL* slope, sheet flow dominates, but velocity increases as the gradient increases downslope, inducing greater erosion. On a smooth *CL* slope, sheet flow is effective, with the steeper gradient upslope yielding higher velocity and greater erosion and the flatter gradient downslope giving lower velocity and promoting deposition.

Curvature along the slope width also affects runoff. An *LV* slope has diverging slope lengths or flow paths, and water tends to disperse. An *LC* slope has converging slope lengths or flow paths, and runoff channels, possibly causing gullies and

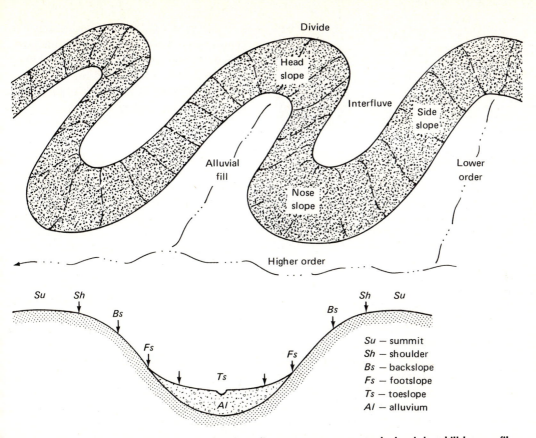

Figure 6.2 The geomorphic components of a slope bounding an open-system watershed and along hillslope profile. From Ruhe and Walker (1968), by permission of Int. Soc. Soil Sci.

forming first-order drainage lines. In group III slopes *(VV, VC, CV,* and *CC),* curvatures along the slope length and the slope width form very complex runoff systems in terms of pattern and velocity. Here, concentration of flow on the hillslope in first-order drainage lines is the beginning of discharge through the drainage net of the open system.

There are other measures that are useful in the study of hillslopes (Savigear, 1956). The *measured length* is the ground distance between any two stations along a slope profile. A *slope segment* is a portion of the profile that differs in gradient from segments above or below it and is generally measured as a multiple of a unit length. A

slope facet is a uniformly inclined surface on a slope which differs distinctly in gradient from those above and below it. (A *scarp* is a steeply inclined slope facet.) A *slope envelope,* an imaginary surface that passes over minor irregularities on the ground surface, makes it possible to ignore microrelief.

The direction of slope is important in analyzing weathering, erosion, and mass movement on hillslopes. The *slope aspect* is the direction of exposure. In northern temperate regions, snow cover persists on north-facing slopes after it has melted on south-facing slopes. South-facing slopes receive direct insolation from the sun, unlike north-facing slopes.

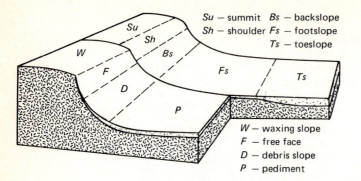

Su — summit   Bs — backslope
Sh — shoulder   Fs — footslope
               Ts — toeslope

W — waxing slope
F — free face
D — debris slope
P — pediment

Figure 6.3 The elements of a "fully developed hill-slope." Foreground from Wood (1942) and L. C. King (1957), by permission of Edinburgh Geol. Soc. Background from Ruhe (1960), by permission of Int. Soc. Soil Sci.

The slope aspect accounts for differences in soil temperature, soil moisture, and vegetative cover. On the Allegheny Plateau in Ohio and beneath the leaf litter under forest, average monthly maximum temperatures on the surface are greater on southwest-facing than northeast-facing hillslopes (Finney et al., 1962). On the litter, temperatures range from 3°F in December to 15°F in June; beneath the litter, from 1°F to 6°F. From May to November soil moisture in percent is greater at depths of 1 to 3 inches and 6 to 9 inches on the northeast than southwest slopes. In May the moisture content on southwest slopes is about 80 percent of that on northeast slopes, but in October it is less than 50 percent.

The climatic effect on vegetation is distinct. A mixed association of oak and hickory is on southwest slopes, but a mixed mesophytic association of beech, tulip tree, maple, cherry, walnut, butternut, basswood, elm, sassafrass, ash, oak, birch, and hickory is on northeast slopes.

On the Cumberland Plateau in Kentucky and Tennessee, average monthly temperatures at a soil depth of 20 inches are greater on south-facing than on north-facing slopes (Franzmeier et al., 1969). The average annual difference is about

1.5°C for upper, middle, and lower hillslope sites. For depths up to 20 inches, field moisture in percent is lower in soils on the south-facing slope. Given the effects of climate and vegetation on soil formation (Chapter 2), one would expect soils to differ according to aspect.

Geomorphic components

In an *open system* (Chapter 5), a stream that incises a valley is part of a drainage net and descends to and joins a higher-order stream. The incised valley of the lower-order stream is closed along the sides and at the head but is open at the mouth. Geomorphic components are as follows: The *head slope* is concave along the slope width at the head of the drainageway, and the slope lengths converge downward to the drainageway (fig. 6.2). *Side slopes* bound the drainageway along its sides and are generally linear along the slope width.

A number of lower-order drainageways join the higher-order member of the net. Between adjacent drainageways is an *interfluve*, and opposed side slopes descend to valley bottoms. At the open end of adjacent drainage basins, a convex *nose slope* curves from one opposed side slope to the other side slope of the interfluve. On the nose slope, slope lengths diverge downward. Three major kinds of overland flow operate: convergent flow on the head slope, parallel flow on side slope, and divergent flow on the nose slope.

On any of these three kinds of slope, other components comprise the *hillslope profile* (fig. 6.2). The highland of the divide or interfluve is the *summit*. Toward the drainageways a slope that is convexly rounded is the *shoulder*. Downward, a *backslope*, which may be linear, descends to a concave *footslope*. Toward the drainageways, with a flattening gradient, is the *toeslope*. The backslope is most susceptible to erosion, the footslope and toeslope to deposition.

Not all of these profile components are on every hillslope. A steep backslope may ascend to a level summit, and the shoulder may be no more than an edge. The angular junction at the edge is common where a cap of caliche, laterite, or gravel is on the summit. The angular junction may be at the base of a backslope where a stream impinges, and footslope and toeslope are absent.

Hillslope profile components fit in the framework of the "fully developed slope" (Wood, 1942; L. C. King, 1957). If the profile (fig. 6.2) is elongated and flattened, a waxing slope, free face, debris or talus slope, and pediment evolve (fig. 6.3). The *waxing slope* is the convex crest of a hill. The *free face* is the outcrop of bedrock or the scarp just below the crest. The *debris slope*, which has detritus derived from the free face, rests against the lower part of the scarp. The *pediment* is a broad concave-upward surface extending away from the debris slope and down to the alluvial plain of an adjacent stream. These "elements" classified by Wood and King are more than hillslope components; they are really parts of erosion surfaces and can occupy considerable area. (See fig. 5.1.)

An alluvial toeslope, to which all other ele-

Figure 6.4 "Fully developed slopes" in New Mexico (a) and Puerto Rico (b). Compare the curvatures and profiles along the interfluve axes (cf. fig. 6.3). The arid-country surface is a pediment because it is arid. The humid-country surface is a peneplain (St. John's) because it is humid.

a

b

Figure 6.5 Cross sections of hills in the Shenandoah Valley, Virginia (a) and in the Altamont end moraine of the Des Moines drift lobe in Iowa (b).

ments descend (fig. 6.3), must be added as an integral part of any erosion-sedimentation system and it is directly applicable to soil landscapes and the soil catenas of Milne (Ruhe, 1960). In a soil catena in areas of residual granite hills in East Africa (Milne, 1936b), dark gray loams form on the hillcrest and shoulder and move downhill across backslope and footslope. Waters wash the footslope and carry finer sediment to the toeslope, forming its clayey floor. The soil catena from high to low ground has gray hill-brow, red-earth, swamp-fringe, and swamp soils.

Similarity or uniformitarianism of hillslopes (L. C. King, 1957) exist to some degree under most climatic conditions (Frye, 1959). Examine the landscape along the Rio Grande in southern New Mexico. An upland summit of limestone gravel which is currently under 7 inches of annual rainfall stands above the other slope elements (fig. 6.4(a), right middle and foreground). A narrow shoulder is held up by cemented caliche or calcrete. Descending to the left is a concave backslope that continuously flattens to footslope and toeslope. These hillslope elements are incised by a drainage net but have a concave upward profile much like the longitudinal profile of a stream.

Now examine the slope profiles along interfluves in Puerto Rico on andesite tuff which are currently under 90 to 95 inches of rainfall [fig. 6.4(b)]. From the small banana grove at the right, a concave backslope descends to a flattening footslope toward the two sheds at middle distance. At the left, a concave backslope descends to a flattening footslope beyond the large shed. Note the similar profiles along the other interfluves in this incised landscape.

Also note the distinct similarity of slope form regardless of climate or underlying material in the two areas. If the biases of "arid" and "humid" geomorphologists are applied, the described New Mexican surface is called a pediment, the described Puerto Rican surface a peneplain (in fact, part of the St. John's peneplain). Herein lies a dilemma, whose solution may be deferred by simply recognizing both composite slope forms as erosion surfaces and awaiting further discussion of the subject.

Internal form of hills

In addition to its external form, which is expressed by the hillslope components, a hill also has an internal form or "structure." Fitting external form to internal form helps in reconstructing the evolution of the landform.

If natural or man-made cuts are available, a cross section of the hill is exposed for study. Examine the interior of a hill in Iowa [fig. 6.5(b)]. A forest bed containing spruce logs is buried 64 feet beneath the hill crest. The wood is radiocarbon dated at $13,900 \pm 400$ years and is in silt and sand between glacial tills. The forest bed is marked in the cross section by the horizontal line in the center about a quarter distance up the cut. Note that left and right hillslopes bevel all beds and are an erosion episode that is younger than all beds within the hill. If this hill were analyzed from the surface form only, it would be just another knob in a glacial moraine.

Examine a cross section of a hill in the Shenandoah Valley of Virginia in the area of "dynamic equilibrium." A Reddish-Brown Lateritic soil is beneath the summit in fine-textured material [fig. 6.5(a)]. Note that the left and right hillslopes bevel the soil, a gravel bed just beneath the soil,

Table 6.1 Average runoff and sediment yield on Almena Silt Loam soil, Wisconsin, 1947-1955

Cover	Runoff (in./yr)		Sediment yield (tons/acre/yr)	
	Rilling	Sheet wash	Rilling	Sheet wash
Corn	2.07	0.95	3.12	0.38
Oats	1.78	0.92	0.90	0.39
Grass	0.82	0.47	0.02	0.02

From Hays and Attoe (1957).

and weathered debris beneath the gravel. Beveling is by the backslope, and it terminates at the hillslope shoulder. Soils on the backslope have little if any development in contrast to the well-developed Reddish-Brown Lateritic soil beneath the summit.

The angular relations, the beveling of beds and summit soil, and the differences between soils on summit and backslope show geologic unconformity. There is a hiatus of considerable magnitude at the shoulder.

The summit soil, because of lateral truncation, formed on a broader ridge crest than currently exists. Stream incision occurred on each side of the hill, which permitted hillslope erosion and beveling of the soil and beds within the hill. At least two geomorphic episodes are separated by the unconformity and its hiatus. Can the occurrence episodes be rejected and be replaced by an explanation of dynamic equilibrium (Hack, 1965)? Such an explanation (Hack, 1960) requires that: (1) A landscape and the processes molding it are a part of an open system in a steady state of balance in which every slope and every form is adjusted to every other. (2) Changes in topographic form take place as equilibrium conditions change (it is not necessary for evolutionary changes ever to occur). (3) Within a single erosional system all elements of the topography are mutually adjusted so that they are downwasting at the same rate. (4) The forms and processes are in a steady state of balance and may be considered independent of

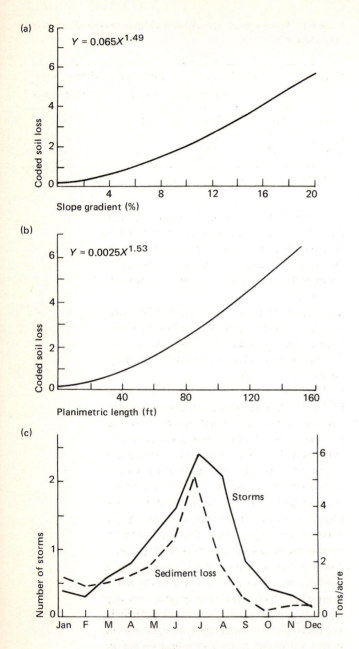

(a)

$Y = 0.065X^{1.49}$

Coded soil loss

Slope gradient (%)

(b)

$Y = 0.0025X^{1.53}$

Coded soil loss

Planimetric length (ft)

(c)

Number of storms

Storms

Sediment loss

Tons/acre

Jan F M A M J J A S O N Dec

Figure 6.6 The relation of sediment or soil loss (erosion) on hillslopes to (a) slope gradient, (b) slope length, and (c) frequency of storms. From Zingg (1940), by permission of Am. Soc. Agric. Engrs., and Barnett and Hendrickson (1960).

time. (5) Differences and characteristics of form are explainable in terms of spatial relations in which geologic patterns are the primary consideration rather than evolutionary development.

Do these requirements explain the hillslope example [fig. 6.5(a)]? Unconformity and hiatus at the hillslope shoulder between summit and backslope show that parts of the same hillslope are not even in adjustment.

Processes on hillslopes

Processes on hillslopes are controlled directly by the force of gravity. After the infiltration capacities of the soils are reached, water runs off downslope as *overland flow* (Horton, 1945), causing erosion and deposition. Water infiltrating the soils on slopes moves downward and laterally, and seepage forces may cause sapping or spalling of material above the seep line. Water infiltrating the soils and substrata may alter the physical properties of the mass, causing its movement downslope.

Erosion and deposition

Hillslope erosion depends on the initial movement of grains and the transport and deposition of eroded material. Factors controlling erosion are the initial resistivity of material, the cover on the hillslope, rainfall intensity, and the velocity and energy of runoff. Overland flow may be unconfined as sheet flow or confined in rills as channel flow.

Initial movement of grains on a hillslope may be caused by direct raindrop impact, by water impact in runoff, or by impact from another particle in transport. The impact of grains requires a barren or sparsely covered surface. Consequently ground cover is a controlling factor of hillslope erosion. Examine Almena Silt Loam soil in Wisconsin, where runoff in inches per year and sediment yield in tons per acre per year were

Table 6.2 Average runoff and sediment yield on Monona and Ida soils, southwestern Iowa, 1964-1969

Cover	Runoff (in./yr)			Sediment yield (tons/acre/yr)		
	Base flow	Surface	Total	Sheet rill	Gully	Total
Brome grass	2.95	1.80	4.75	0.3	0.5	0.8
Corn (contoured)	2.57	5.18	7.70	30.0	8.0	38.0
Corn (terraced)	6.63	0.77	7.40	0.9	0.0	0.9

From Saxton et al. (1971), by permission of Am. Soc. Civil Eng.

measured from 1947 through 1955 (table 6.1). The average annual rainfall was 27.5 inches. The crops of corn, oats, and grass have increasing density of cover. Note that runoff and sediment yield from rilling and sheet wash decrease with increase in density of cover.

The shape of the hillslope in combination with cover also controls runoff and erosion. On loess in southwestern Iowa under 33 inches of annual rainfall, total annual runoff is less under grass than under contoured or terraced corn (table 6.2). Total runoff is about the same from contoured or terraced corn, but surface runoff is much less from terraced corn. Water is held on the terraced slopes, which permits infiltration and contribution to base flow (Chapter 3) and decreases sediment yield (erosion).

Intensity of rainfall also effects hillslope erosion. Sediment loss is related to the frequency of storms throughout a year [fig. 6.6(c)] where a storm has half an inch or more of rain in half an hour (Barnett and Hendrickson, 1960). On Cecil soils on the Piedmont in Georgia from 1940 to 1960, about 11 storms a year causing only 25 percent of the total rainfall caused 86 percent of the erosion and 56 percent of the runoff.

The geometry of the hillslope also controls runoff and erosion. Soil loss in weight units is directly related to slope gradient and also to planimetric slope length, as expressed by the equation $Y = aX^b$ [fig. 6.6(a), (b)]. Combining these two relations gives (Zingg, 1940):

$$X = kS^m L^n$$

where X is soil loss in weight units; S, slope gradient in percent; and L, planimetric length in feet. Rainfall intensity enters the erosion equation (Musgrave, 1947) as:

$$E = P_{30}^{1.75} S^{1.35} L^{0.35}$$

where P is the maximum rainfall in inches in a 30-minute period on a 2-year frequency basis.

Other factors are introduced in a "universal soil-loss equation" (Wischmeier and Smith, 1965):

$$A = RKLSCP$$

where R is a rainfall factor measuring the erosive force of a specific rainfall; K, a soil-erodibility factor, or a measure of the resistance to erosion of a given material; L and S, length and gradient factors; C, a cropping-management or cover factor; and P, an erosion-control practice factor (showing whether the slope is terraced, contoured, and so forth). The factors are determined from experimental plot studies, and analytical data defining each of the factors are used in the equation.

These empirical relations transpose to geomorphic principles. (1) The materials of the hillslope must be erodible. (2) Either cover must not be dense or it must be breached so that erosion may occur. (3) The severity of erosion depends on the gradient, length, and shape of the slope and the intensity and duration of rainfall.

Rilling, a more effective erosion agent than sheet wash (table 6.1), is a dominant process in hillslope reduction and involves cross grading and micropiracy (Horton, 1945). A hillslope is

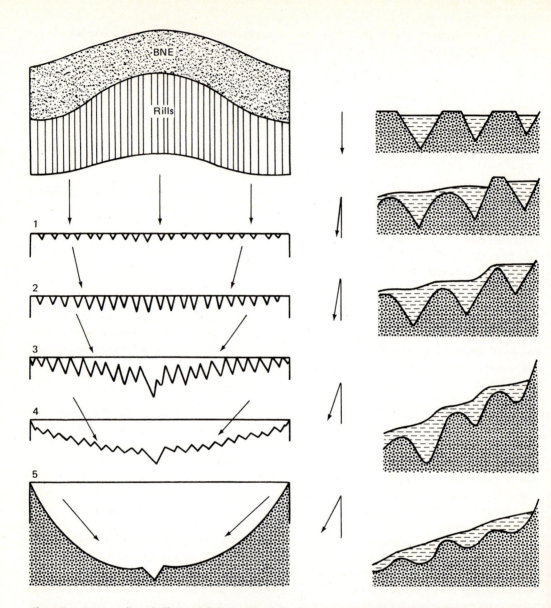

Figure 6.7 Cross-grading of rills on a hillslope with a belt of no erosion (BNE) and a belt of rills in plan view. The sequence of cross-sectional profiles downward in the left-hand column transforms the initial rills to a main drainageway in stage 3 and a hillslope valley in stage 5. The right-hand column shows a detailed sequence of cross-grading of rills downward. Arrows indicate the resultant direction of overland flow. From Horton (1945), by permission of Geol. Soc. Am.

rilled below a Horton "belt of no erosion" (fig. 6.7). Small uniform, parallel channels are closely spaced on a uniformly sloping surface (fig. 6.8). During intense runoff, a divide between two rills is broken by caving, by undercutting by a deeper rill, or by overflow. Water is diverted from the higher to the lower rill in lateral micropiracy. The divide is obliterated by cross-grading, and a sequence of events (fig. 6.7, 1 to 5) may develop a gully (3, 4) and ultimately a first-order drainageway (5).

The Horton "belt of no erosion" is delineated by a "critical distance" from the watershed divide to the point on the slope where the rills begin. According to the Zingg equation ($X = kS^mL^n$) some slope length is required before erosion begins. A critical distance may be required to initiate rilling, but once it begins, headward extension proceeds rapidly to the point of overflow from a higher level (fig. 6.8).

Cross grading causes incision of a number of parallel sideslope valleys which tend to reduce and flatten the hillslope [fig. 6.8(c)]. This process, which is termed "gully gravure" (Bryan, 1940), has been used to explain slope retreat in the White Mountains of California (Beaty, 1959).

Erosion and deposition on hillslopes are important processes in open and closed systems (Chapter 5). Any analysis of erosion and deposition must include the hillslopes. In a closed basin (Walker, 1966), hillslope erosion is estimated by comparing the volume of sediment in the basin to the area of source on the hillslopes. Assuming uniform stripping on the hillslope, volume divided by area gives the thickness removed. Radiocarbon dating of basin sediments (figs. 5.6, 5.7) permit determination of rates of erosion.

Note the differences in thicknesses eroded and erosion rates during episodes marked by basin organic and mineral sediments (table 6.3). The rates of erosion for organic sediments are less than those for mineral sediments, so organic sediments represent episodes of relative slope stability, and mineral sediments represent episodes

of relative slope instability. The system is not steady-state, but is episodic with varying rates.

Hillslopes may have very systematic sedimentation systems that are attributable to overland flow. For example, a hillslope on glacial till in Iowa descends to a closed basin (Walker, 1966). Note the systematic changes from summit to toeslope in the thickness of hillslope sediment, content of gravel and clay particles, and geometric mean particle size (fig. 6.9). The thickness of the surficial sediment increases; gravel decreases; clay increases; and geometric mean size decreases. These relations are readily expressed by equations describing how systematic deposition has been (fig. 6.10).

Another hillslope formed on glacial till has a large interbedded sand and gravel lens (Kleiss, 1970). This lithologic interruption (fig. 6.11) causes a dual sedimentation system on the upper and lower parts of the slope. The geometric mean particle size decreases from the summit across the shoulder to the backslope, where a distinct increase occurs at the contact with the sand lens (fig. 6.12). A second particle-size distribution decreases progressively downslope. Note the opposite relation in clay content as it increases, decreases, and again increases. These systems of sorting are clearly evidence of former overland flow on the hillslope.

Mass movement

Water on a hillslope not only erodes, transports debris, and deposits sediment, but it infiltrates the mass and may alter its physical properties. For various states of matter, the mass may move downslope under force of gravity.

Soil dispersed in water has all the characteristics of a liquid. It flows, and its specific gravity can be measured with a hydrometer. This mixture is in the *liquid state*. If water is partially removed from this mixture but all voids are filled, the state is called *viscous*. The mixture will deform under its own weight, but its specific gravity cannot be

a

b

c

Figure 6.8 (a) A rilled face of a road cut on Cary glacial till, buried soil, and lower glacial till along a freeway interchange in Des Moines, Iowa. Note that the rills extend to the top of the slope. (b) Valley-slope and bottom gullies in loess and alluvium in Harrison County, Iowa. Note that the valley-slope gullies extend to the summit. From Daniels and Jordan (1966). (c) Gully gravure on a hillslope in southern New Mexico.

measured with a hydrometer. The boundary between the two states is the *liquid limit,* and cohesion is essentially zero.

With further reduction of moisture, the mixture shifts to the *plastic state.* Pore spaces are no longer completely filled with water, and capillary forces tend to bind soil particles. In the plastic state an external force must be applied to deform a mass if it is to remain deformed after the force is released. The *plastic limit* is the smallest water content at which a soil is plastic. With continued removal of water, the mass shrinks to a point at which further reduction in moisture causes no shrinkage (fig. 2.1). The *shrinkage limit* is the moisture content below which the soil does not shrink. Below it the mass reaches the *solid state.*

These four states, which occur naturally, control mass movement on hillslopes. Beneath a hillslope in unconsolidated sediment, moisture contents vary, and the state of the solid-liquid mix can be complicated. Note differences in gravimetric moisture and water saturation across a drainageway in Iowa (fig. 6.13). The water is perched in loess above paleosols and glacial till (Vreeken, 1968). *Gravimetric moisture* is the water weight expressed as percent of the dry weight of the sediment. *Volumetric moisture*, the percent water volume in the total solid-liquid mix, measures the state of water saturation. Under each hill summit moisture is relatively high; downward it decreases and then increases in the subsurface to a saturated zone in the alluvium of the drainageway. The solid-liquid zoning gives different mechanical properties stratigraphically and complicates the strength of materials under the hillslopes.

Earth materials have inherent properties that tend to resist deformation when force is applied (Taylor, 1948). *Shearing strength* is the resistance of a material to shear along an internal plane caused by an applied external force. *Shearing stress* is the applied force which causes failure. Just prior to failure, shearing strength equals shearing stress and is the *maximum strength* of the

Table 6.3 Hillslope erosion around closed basins in central Iowa

Bog strata	Thickness (ft)	Rate (in./1000 yr)
Colo bog		
Upper muck	0.2	0.72
Upper silt	1.7	4.08
Lower muck	0.03	0.07
Lower silt	0.9	5.40
	2.8 (total)	2.57 (average)
Jewell bog		
Upper muck	0.2	1.08
Upper silt	5.3	8.76
Lower muck	0.5	6.00
Lower silt	0.6	7.20
	6.6 (total)	5.76 (average)

From Walker (1966).

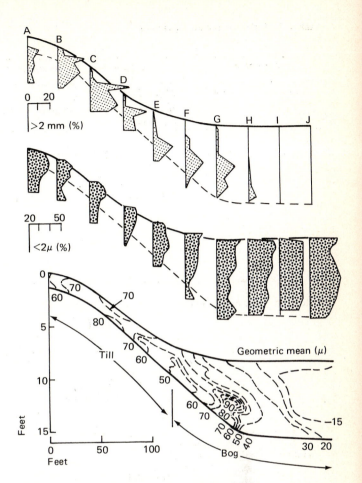

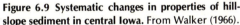

Figure 6.9 Systematic changes in properties of hillslope sediment in central Iowa. From Walker (1966).

material. When shearing stress exceeds shearing strength, failure occurs.

Shearing strength depends on the cohesion and internal friction in a material. *Cohesion* is the molecular attraction that binds particles together. *Internal friction* is the force required to overcome all frictional resistance and cause slip along a plane through a mass. With reference to force diagrams (fig. 6.14), a body is at rest on a flat horizontal surface (a) with a total vertical force N and an opposing force R. As long as no other force is applied, there is no friction, although friction (F) is available if other force is applied. If a horizontal force S is applied (b), the resultant P of vertical force N and horizontal force S is at angle a to the normal force. An opposing force becomes active at angle a' to R. Angle a' is less than angle i which subtends all available friction, now $F + (F)$, and slip does not occur along the plane because $F + (F)$ is dominant. If enough force S is applied to equal full friction F, angle a becomes equal to angle i and slip is imminent to the left (c). Angle i is the *internal friction angle*. The tangent of i is F/R

or S/N or μ, which is the *coefficient of friction*, and $S = N \tan i$ is the *internal friction force*.

The amount of available friction depends on the applied force and the friction angle. When the angle of applied force is less than the friction angle, slip is not possible, but when the angle of force equals the friction angle, slip is imminent. If weight and reaction to weight are the only forces acting on a body on a slope [fig. 6.14(d)], slip will occur when a *critical slope g* is reached and when the angle of that slope equals the internal friction angle.

Table 6.4 Summary of mass movement

Movement		Composition of mass and process			Favoring conditions
Kind	*Rate*	*Material dry or with minor ice or water*	*Material and water*	*Material and ice*	
Creep	Very slow	Soil creep	Rockcreep Talus creep	Solifluction	Unconsolidated sediment or structurally modified rock. Bedded or alternate resistant and weak beds. Rock broken by fractures, joints, etc. Slight to steep slopes. High daily and annual temperature ranges; high frequency of freeze and thaw; alternate abundant rainfall and dry periods. Balance of vegetation to inhibit runoff but not to anchor movable mass.
Flowage	Slow to rapid		Earthflow Mudflow Debris avalanche	Debris avalanche	Unconsolidated materials, weathering products; poorly consolidated rock. Alternate permeable and impermeable layers; fine-textured sediment on bedrock. Beds dipping from slight to steeper angles; beds fractured to induce water in cracks. Scarps and steep slopes well gullied. Alpine, humid temperate, semiarid climate. Absence of good vegetative cover such as forest.
Sliding	Slow to very rapid	Slump Debris slide Debris fall	Rockslide Rockfall		Inherently weak, poorly cemented rocks; unconsolidated sediments. One or more massive beds overlying weak beds; presence of one or more permeable beds; alternate competent and incompetent layers. Steep or moderate dips of rock structures; badly fractured rock; internal deforming stress unrelieved; undrained lenses of porous material. Scarps or steep slopes. Lack of retaining vegetation.
Subsidence	Slow to very rapid		Subsidence		Soluble rocks; fluent clays or quicksand; unconsolidated sediments or poorly lithified rocks; materials rich in organic matter, water, or oil. Permeable unconsolidated beds over fluent layers. Rocks crushed, fractured, faulted, jointed inducing good water circulation. Level or gently sloping surface.

Compiled and modified from Sharpe (1938), by permission.

Causes

Wedging and prying: by plant roots; swaying of trees and bushes in wind; expansion of freezing water and hydrostatic pressure of water in joints and cracks; diurnal, annual, irregular expansion due to heating; expansion due to wetting; animal activity. Filling and closing of cracks and voids caused by: burrowing of animals; decay of plant roots and other organic matter; gullying or undercutting by streams; removal of soluble rocks and minerals; erosion of fine particles by sheet wash and rills; downslope mass movement; shrinkage due to drying or cooling. Increase in load: addition of material upslope; rainfall, snow, or ice; traffic of vehicles or animals; tectonic, meteorologic or animal disturbance.

Reduction in internal friction due to excessive amounts of water in mass. May start as slide; causes similar to landslides.

Removal of support: oversteepening of natural or artificial slopes by erosion; outflow, compaction, softening, burning out, solution, chemical alteration of subjacent layer; disappearance of buttress against slope such as ice front.
Overloading: by other mass-movement processes; by rain, snow, ice, and saturation; overburden in excavation.
Reduction of internal friction and cohesion: by surface and ground water, oil seeps, chemical alteration by weathering.
Wedging and prying: as in creep.
Earth movement: produced by earthquakes; storms, traffic of vehicles and animals; drilling, blasting, gunfire; earth strains due to temperature and atmospheric pressure and tidal pull.

Removal of support of subjacent layers: by solution or chemical alteration; by outflow of fluent material; by natural or artificial excavation; by compaction caused by natural or artificial overloading, by reduction of internal friction, by desiccation.
Earth movement: by warping; by natural or artificially induced vibrations.
Overloading: natural or artificial.

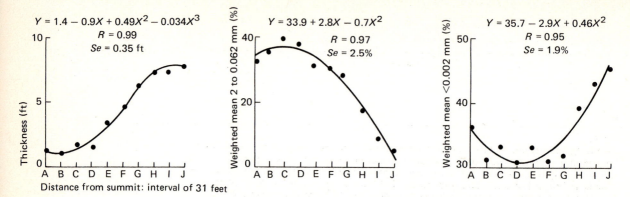

Figure 6.10 Curve fitting and empirical equations expressing the relationship of properties of hillslope sediments. (Cf. fig. 6.9) From Walker (1966).

The *Coulomb equation* expresses the relation of shearing strength S to cohesion and internal friction:

$$S = C + N \tan i$$

where C is cohesion in pounds per square foot; N, the normal force in pounds per square foot; and i, the internal friction angle in degrees. In sediments clay furnishes cohesion, and sand and silt supply internal friction. Cohesion is independent of load, but internal friction is dependent upon load or applied force.

Various states of mass and forces on a hillslope cause various kinds of mass movement that are classified one way by geologists (table 6.4) and another way by engineers (table 6.5). The viscous state of a mass is caused by thorough water saturation following excessive rainfall. In the Allegheny Plateau, earth slips occur on fairly steep, unstable slopes, with slowly permeable subsoil and excessive surface and subsurface water (Patton, 1956). In central Virginia debris avalanches followed torrential rains on August 19 and 20, 1969 (Williams and Guy, 1971). Remnants of Hurricane Camille dumped 27-28 inches of rain in 8 hours. Rapid downhill flow of earth, rock, vegetation, and water left hillslope scars 200 to 800 feet long, 25 to 75 feet wide, and 1 to 3 feet deep. In southern California in 1941, Wrightwood mudflow moved almost 15 miles in a descent of 5000 feet (Sharp and Nobles, 1953). Water content from melt water was 25-30 percent by weight, and the flow attained velocities of 14 feet per second.

The plastic state of a mass usually results in creep or slide movement. Since creep is imperceptible, measurement is recorded by movement of surface stakes, pins, or columns in the ground, by movement of tilt bars or deformation of tubes, or by movement of plates (Selby, 1966). Shear takes place, and one part of the measuring device is moved farther from a bench mark than another part.

Creep, which may be seasonal or continuous (Schumm, 1964), is caused by wetting and drying, freezing and thawing, and animals' burrowing (Kirkby, 1967). Upon wetting, colloidal components swell, and expansion is perpendicular to the slope. Upon drying, colloids shrink, and contraction is along a resultant between the normal and vertical vectors causing net downslope movement (Young, 1960). Frost action produces the same effect.

The rate of movement on hillslopes is proportional to the gradient (Culling, 1963), to the sine of the slope angle, or to that component of gravitational force acting parallel to the hillslope (Schumm, 1967b). Measurement of creep is a long-term process and requires patience.

Slides are more readily analyzed by the methods of soil mechanics (Taylor, 1948). If one suspects sliding on a hillslope, one collects undisturbed cores from the mass and subjects them to various normal loads and shearing stresses in the laboratory. At a given normal load, failure occurs at a given shearing stress. Values of shearing stress (Y) plotted against normal load (X) yield a straight line expressed as $Y = a + bX$. Intercept a on the Y axis measures cohesion, and the slope of the line b is the internal friction angle. The unit weight of soil and water is also measured in the laboratory from core samples. Values of cohesion, internal friction angle, and unit weight are used to analyze the slide.

From the field measurements draw the hillslope and subsurface mass to scale [fig. 6.15(a)] and analyze the drawing using the *method of slices* (Fellenius, 1936). Select a trial subsurface arc. Divide the cross section into equal vertical slices, using unit width for volume. Multiply unit weight by volume for each column, giving W. Draw a vector W vertically to scale for each column. Resolve W into its components of the nor-

Figure 6.11 A cross section of a hillslope in northeastern Iowa and equations for describing the profile. From Kleiss (1970), by permission Soil Sci. Soc. Am.

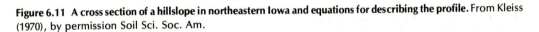

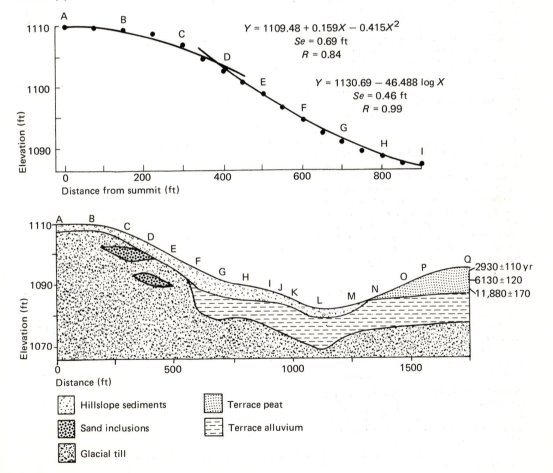

$$Y = 1109.48 + 0.159X - 0.415X^2$$
$$Se = 0.69 \text{ ft}$$
$$R = 0.84$$

$$Y = 1130.69 - 46.488 \log X$$
$$Se = 0.46 \text{ ft}$$
$$R = 0.99$$

Elevation (ft)

Distance from summit (ft)

2930 ± 110 yr
6130 ± 120
11,880 ± 170

Elevation (ft)

Distance (ft)

Hillslope sediments

Terrace peat

Sand inclusions

Terrace alluvium

Glacial till

Table 6.5 Classification of mass movement

Type of movement	Bedrock	Soils	
		Nonplastic	*Plastic*
FALLS: Mass in motion travels most of the distance through the air. Includes free fall, movement by leaps and bounds, and rolling of rock and debris fragments without much interaction of one fragment with another.	Rockfall	Soilfall	
SLIDES: Movement caused by finite shear failure along one or several surfaces which are visible or whose presence may be inferred.			
Material in motion not greatly deformed: Moving mass consists of one or two units. Maximum dimension of units is greater than displacement between them. Movement may be controlled by surfaces of weakness such as faults, bedding planes, or joints.	Block-slide Slump		Slump
Slump: Movement only along internal slip surfaces, which are usually concave upward. Backward tilting of units is common.			
Block-glide: Movement of a single unit out and down along a more or less planar surface of weakness, generally a bedding plane. Block may glide far on original ground surface.			
Material in motion greatly deformed or consists of many independent units; movement frequently is structurally controlled by surfaces of weakness such as faults, joints, bedding planes, variations in shear strength between bedded deposits, or by contact between bedrock and overburden. Maximum dimension of units is equal to or less than displacement of center of gravity of whole mass. Movement may progress beyond original slip surface so that parts of mass slide over original ground surface.	Rockslide	Debris slide	
FLOWS: Movement within displaced mass such that the form taken by moving material or the apparent velocities and displacements resemble those of viscous fluids. Slip surfaces within moving material are usually not visible or short-lived. Boundary between moving mass and material in place may be sharp or a zone of distributed shear.	Rock flow (Dry)	Sand flow	
		Debris avalanche	Earthflow
	(Wet)	Debris flow	Mudflow
COMPLEX LANDSLIDES: Movement is by combination of one or more other types of movement. One type generally dominates others at certain areas in the slide or at a time during slide.			

From Horner (1953).

116   HILLSLOPES

mal N and the tangential Wt in the direction in which it acts. Sum the Wts algebraically. Positive values act outward from the hillslope which causes sliding, and negative values act inward toward the hill, resisting sliding. Sum the N values.

Now solve graphically (Taylor, 1948). Along a horizontal base line, lay off $\sum N$ to scale [fig. 6.15(b)]. From a point of origin on the base line, construct the internal friction angle i, which was determined in the laboratory. From the base line construct a vertical line from point $\sum N$ and intercept the internal friction angle i. From the base line to the intercept equals $\sum N \tan i$ or the sum of internal friction. On the vertical axis add to scale total cohesion Ct, which was determined in the laboratory for samples along the trial arc. Then:

shear strength = $\sum N \tan i + Ct$

shear stress = $\sum Wt$

where $\sum Wt$ is scaled along the vertical from the base line.

So:

$$\frac{\text{shear strength } S}{\text{shear stress } s} = f$$

If $f > 1$ there is no failure of slide, but if $f < 1$, failure is imminent. The foregoing example is one of the many methods for analyzing slides (Taylor, 1948).

Mass movement changes the shape of a hillslope. Dominance of creep over rainwash erosion causes convex slopes (Schumm, 1956c). Flowage and slides form theaterlike hollows on hillslopes that are similar in form to first-order drainageways. Examine a slump in a 45° slope railroad cut in loess and glacial till (fig. 6.16). The mass slid many feet along the vertical headwall of a critical surface with the toe protruding from the cut. In 6 months the slide collapsed into a jumbled mass with many individual slides and flows. The end form 2 years later was a spoon-shaped hollow, causing slope retreat.

Some landslides are huge features. Blackhawk

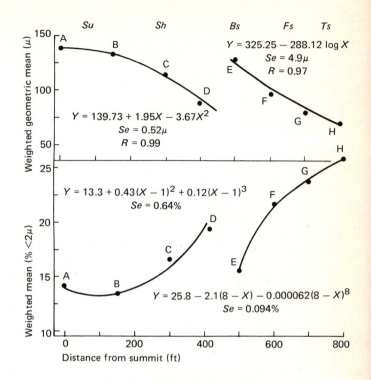

Figure 6.12 Empirical descriptions of sediment properties on the hillslope. Note the effect of the sand lens at site D (fig. 6.11) causing a dual-system on the slope. From Kleiss (1970), by permission of Soil Sci. Soc. Am.

landslide in the San Bernardino Mountains in California produced a lobe 30 to 100 feet thick, 2 miles wide, and 5 miles long (Shreve, 1968). A slide of this size apparently begins as an ordinary rockslide, but at a sudden steepening gradient, it leaves the ground, overriding, and compressing a cushion of air. Friction is reduced, and the mass readily moves across gentler slopes.

Solifluction lobes and terracettes have concave upward longitudinal profiles (fig. 6.17). Scars remaining in the source area from which flow occurred have a spoon-shaped depression form. Where bedrock is uncovered by such flowage, *altiplanation terraces* form (Péwé, 1970). Mass movement processes cause *slope retreat* and *slope reduction*—retreat is the movement of the

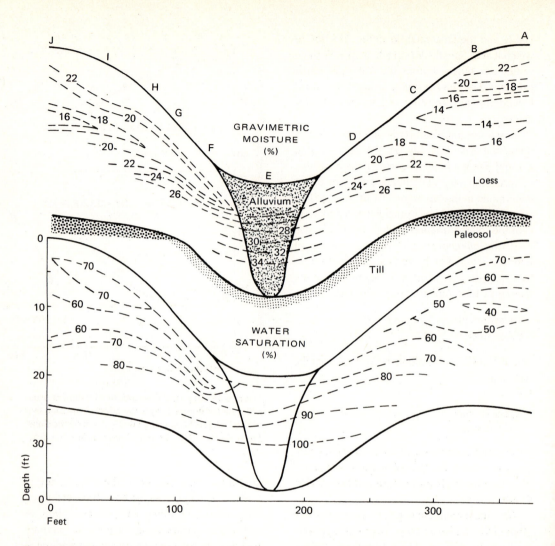

Figure 6.13 Moisture content and water saturation in loess over paleosols and glacial till in central Iowa. From Vre-eken (1968).

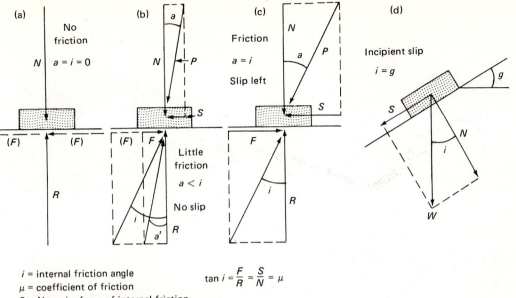

i = internal friction angle
μ = coefficient of friction
$S = N \tan i$ = force of internal friction

$$\tan i = \frac{F}{R} = \frac{S}{N} = \mu$$

Figure 6.14 Mechanical principles of shearing stress and internal friction.

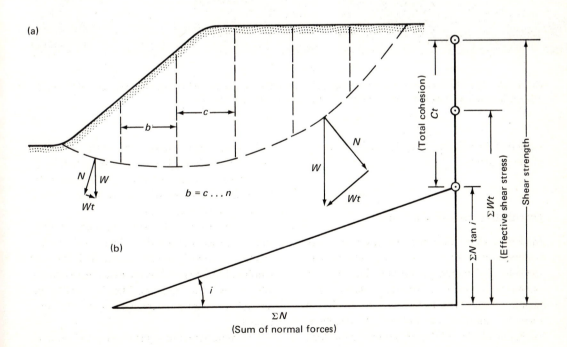

Figure 6.15 Analysis of a landslide on a hillslope by the method of slices (a) and by graphic solution of mechanics (b).

September 1953

May 1955

April 1954

July 1957

Figure 6.16 Formation of a slump mass in a railroad cut in loess and glacial till in Iowa. Note the vertical headwall and toe protrusion along the critical surface, rotation, collapse, and flowage.

slope into the hill, and reduction is the flattening of the gradient.

Quantitative analysis of hillslopes

Modern work on hillslopes requires quantitative methods; one can choose from mathematical-theoretical, dynamic-geomorphologic, or morphometric-statistical techniques (Bakker and Strahler, 1956). The first technique involves a massive infusion of differential equations geometrically describing the profile of a hillslope

(Lehmann, 1933; Bakker and LeHeux, 1946, 1947, 1950, 1952; Looman, 1956). There is a question as to whether this kind of treatment offers insight into processes such as erosion, deposition, or mass movement.

Dynamic-geomorphologic treatment of hillslopes is process-oriented (Strahler, 1952). Mathematical models based on rational deduction or empirical analysis of observed data are used to relate energy, mass, and time. The slice method of analysis of mass movement illustrates this approach.

The morphometric-statistical method permits

Figure 6.17 Solifluction lobes or terracettes along Denali Road in the Maclaren River Valley, Alaska. Note the slope flattening.

analysis of hillslopes as two-dimensional profiles or as the external rind of multidimensional solid geometric forms (Ruhe and Walker, 1968). The first is easier, and it is facilitated by curve fitting. The most widely used method of curve fitting, by least squares, is advantageous because of statistical devices that are available for testing goodness of fit and significance of relations (figs. 6.10, 6.11, 6.12).

A simpler way to treat hillslopes is by frequency distributions of slope classes (Ruhe, 1950; Strahler, 1950; Speight, 1971), which quantify abundance of hillslopes of specific ranges of gradients. The reasons for differences in the distributions of one area versus another may be differences in age, lithology, relief, vegetation, or climate. All of these techniques aid in organizing, describing, and testing data which can be arranged in models and systems.

Soils on hillslopes

Soils, which are part of the hillslope models and systems, respond to geomorphic processes that are or have been operative on landforms.

Examine an interfluve in Iowa where loess overlies paleosols and glacial till (fig. 6.18). Some of the hillslopes have a loess mantle, but others are erosional, and loess is stripped, which exposes glacial till. There is side-valley sediment in first- and second-order drainageways on the hillslopes. The soil pattern (c) reflects the kinds of materials, the exposure of paleosols (b), and the topography (a).

A *normal soil* is one whose profile is in equilibrium or nearly in equilibrium with its environment, which has developed under good but not excessive drainage from parent material of mixed mineralogical, physical, and chemical composi-

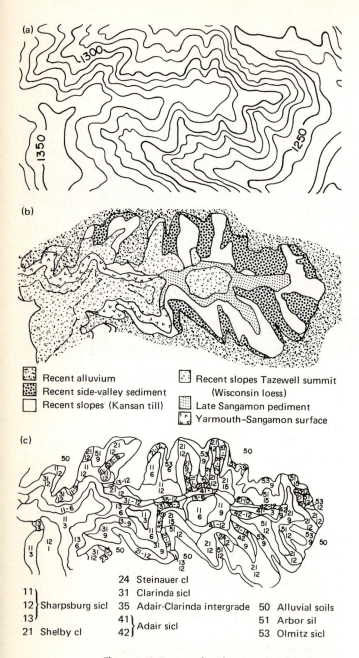

(a)

1300
1350
1250

(b)

Recent alluvium
Recent side-valley sediment
Recent slopes (Kansan till)

Recent slopes Tazewell summit (Wisconsin loess)
Late Sangamon pediment
Yarmouth–Sangamon surface

(c)

24 Steinauer cl
11
12 } Sharpsburg sicl   31 Clarinda sicl
13
21 Shelby cl   35 Adair-Clarinda intergrade 50 Alluvial soils
                      41 }   51 Arbor sil
                      42 } Adair sicl   53 Olmitz sicl

Figure 6.18 Topography of an area in Adair County, Iowa, where loess overlies paleosols and glacial till (a). Loess distribution, exposure of paleosols and till, and side-valley sediment fit the topography (b). Soils (c) in turn fit (b) and (a). From Ruhe and Daniels (1958), by permission of Soil Sci. Soc. Am.

tion, and which expresses the full effects of the forces of the climate and living matter (Soil Survey Staff, 1951). Another requirement for the lengthily defined normal soil is a topographic position of "normal relief" on sloping uplands with medium runoff where, under native vegetation, "erosion removes materials as the solum deepens, thus bringing relatively new minerals into the soil from beneath" (Soil Survey Staff, 1951). So, briefly stated, the normal soil is on hillslopes, and for proper understanding, it must be fitted within the model of summit to toeslope.

Many things about the normal soil are difficult to measure—equilibrium, the full effects of forces of the climate and organisms, and the balance between surface erosion and solum deepening —but the soil-hillslope system can be analyzed. On the hillslope in Iowa mentioned earlier (Walker, 1966), input of climate and organisms is translated as organic carbon content of the A horizons (fig. 6.9). Depth to the thickness of the layer which is more than 2 percent organic carbon (C) is related to distance (D) from summit, as expressed by:

$$C = 7.49 + 1.0D + 0.26D^2$$

The thickness of the A horizon progressively increases from summit to shoulder, backslope, footslope, and toeslope.

On the dual hillslope system in Iowa (fig. 6.11), weighted percent organic carbon (C) in the A horizon relates to the distance from summit (D), as expressed by two equations (Kleiss, 1970):

$$C = 0.91 + 0.071D \text{ (from summit to backslope)}$$

$$C = 2.61 + 0.762D \text{ (from backslope to toeslope)}$$

Thickness of the A horizon (T), expressed as depth to less than 1 percent organic carbon, relates to the distance from summit (D), as expressed by these two equations:

$$T = 9.0 + 0.67D - 0.444D^2 \text{ (from summit to backslope)}$$

$T = 10 + 0.00019D^5$ (from backslope to toe-slope)

Although interrupted downslope, organic carbon increases from the summit to the footslope (Kleiss, 1970).

These organic carbon distributions not only express another part of the sedimentation systems on the hillslopes but they also reflect changing climatic patterns. The lower parts of the hillslopes are wetter than the higher parts, and organic matter production is therefore greater.

Both hillslope studies are excellent examples of the soil catena. In the first, erosion and deposition occurred on one material and were followed by pedogenic processes. In the second, the processes were affected by two materials. These models lead to better understanding of the soils, and any one of the soils is normal within the sequence. There is no need to extract one hillslope soil as normal, implying that all others in the sequence are abnormal.

Deposition on the footslope and toeslope creates *cumulic soils*. As materials accrete, they become incorporated in the upper parts of previously formed soil horizons (fig. 6.19). Pedogenic processes then alter and incorporate them in the pre-existing horizons. The soil profile builds upward and can become exceptionally thick.

Soil moisture also increases at the footslope and toeslope. Note greater saturation at shallower depths than at higher positions on the hillslope (fig. 6.13). Lower-lying members of the soil catena and the cumulic soils have poorer internal drainage. A moister soil environment causes more abundant plant growth, production of organic matter, and production of organic acids. Poor aeration with abundant organic acids produces reducing conditions and gleying of soils.

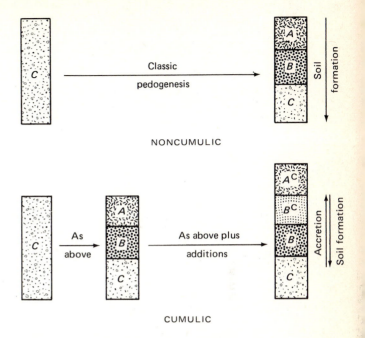

Figure 6.19 Classic pedogenesis and cumulic-soil processes.

An erosion surface is a land surface shaped by the wearing action of water, ice, wind, and other land and atmospheric agents. Since the major agent shaping the land is running water, however, the term "erosion surface" usually refers to a land surface modeled by action of running water.

Identification of erosion surfaces

The first problem in working with an erosion surface out-of-doors is to identify it by its relation (1) to rock and sediment beneath it, (2) in space, and (3) to sediments that lie on it.

Substrata criteria

An erosion surface cuts across beds, materials of different erodibility, and rock and geologic structures. The angular relation of a surface to subjacent beds, materials, and structures is proof that the beds have been cut. On dipping or folded strata the surface may cross without interruption of gradient (fig. 7.1), or the gradient may change at the contact of beds of different resistance, causing descent of a slope facet from the more resistant to the less resistant bed.

To prove that an erosion surface exists where underlying beds are horizontal or have a low dip, there must be evidence that a part of a bed or more has been stripped. This may be difficult to find where the descending gradient of the erosion surface almost parallels the dip of the beds. Erosion surfaces cut across low-dipping basalts in

7

Erosion Surfaces

a

b

c

Figure 7.1 (a) Crosscutting of dipping Devonian beds by an erosion surface near Bonn, West Germany. (b) Crosscutting of steeply dipping quartzite near Tala, Ituri, Zaire. (c) A concave upward erosion surface profile on basalt flows, Oahu, Hawaii. Note the crosscutting of beds on the backslope.

Oahu, Hawaii, and the maximum beveling of beds is beneath the backslope (fig. 7.1).

Spatial relations

Erosion surfaces occupy interfluves between streams in a drainage net and rise to the divide of the drainage basin. Inspect an erosion surface cut in rhyolite on the flank of the Dona Ana dome in New Mexico [fig. 7.2(a)]. The main stream of the watershed crosses from left to right in the foreground and ascends to the mountains in the left background. Tributaries ascend in parallel to a divide in the right background. Examine the interfluves between the parallel tributaries and visualize the longitudinal profiles along the axes of the interfluves. They rise, first gently and then more steeply to the divide in the right background. These profiles are like the concave upward longitudinal profile of a stream.

Now examine the profiles across the interfluves at right angles to the longitudinal axes. They are slightly convex upward, as is shown by the lines of vegetation on the nearly barren bedrock surface.

This kind of erosion surface is a *rock pediment,* common in the deserts of the southwestern United States (Tuan, 1962; Ruhe, 1967), and the shape of the longitudinal profile is characteristic (Mammerickx, 1964; Hadley, 1967). A cross profile from divide to divide in a watershed is scoop-shaped (Bryan, 1935) or concave, which is a result of coalescence of descending tributaries with the master stream at a lower altitude (Gilluly, 1937). Compare this rock pediment to the "fully developed slopes" on gravel and volcanic tuff in arid and humid country (fig. 6.4), and note the similarity of shapes and curvatures within a watershed or drainage basin.

Surficial sediments

Erosion surfaces commonly have surficial sediments lying on them (fig. 7.3). Some of these de-

a

b

Figure 7.2 (a) A rock pediment on rhyolite west of the Dona Ana Mountains, New Mexico. Note the concave upward longitudinal profile along the interfluves. Note the convex upward profile across the interfluves. (Cf. fig. 6.4) (b) Stepped surfaces bordering Rio Grande near Socorro, New Mexico. From Ruhe (1967), by permission of N. Mex. Bur. Mines and Miner. Resour.

posits are fluvial (Trowbridge, 1921), but others are hillslope sediments that were deposited during reduction of the interfluves. A common feature marking the erosion surface is an *erosion pavement,* a layer of gravel or coarse particles left as a *lag concentrate* after finer particles were removed. This layer appears in a vertical section as a *stone line* (fig. 7.3) and has been recognized in soils for many decades (Ruhe, 1959). In roadcuts on the Piedmont in the southeastern United States, a line of angular and subangular fragments parallels the ground at a depth of several feet and sharply delineates the overlying material from underlying bedrock (Sharpe, 1938; Ireland, Sharpe, and Eargle, 1939; Parizek and Woodruff, 1957). There are similar features on erosion surfaces on igneous and metamorphic rock in Africa (Ruhe, 1956a) and on erosion surfaces on glacial till in Iowa (Ruhe, 1956b).

A stone-line surface usually differs topographi-

Figure 7.3 (a) Rhyolite-gravel surficial sediment on a rock pediment on rhyolite near the Dona Ana Mountains. (b) A stone line in paleosol in glacial till in Iowa. (c) The supply of vein quartz to a stone line in weathered granite near Rona, Ituri, Zaire. (d) A stone line in weathered granite near Irumu, Ituri. Note the surficial sediment above the stone line and the rounding of the stone-line gravel. From Ruhe (1959), by permission. © Williams & Wilkins Co., Baltimore.

cally from the present land surface. On the Piedmont in Georgia, stone lines are essentially horizontal in cross sections that cut present hillslopes. In gully exposures that parallel present slopes, stone lines deviate from the horizontal but have less gradient than the present surface. The stone-line surface is truncated by present hillslopes (Parizek and Woodruff, 1957).

Along a longitudinal axis of interfluves on glacial till in Iowa, a stone-line surface emerges from beneath thick alluvium and rises up concave upward profiles from the major stream to the drainage-basin divide. Across the interfluve, the stone-line surface emerges from beneath side-valley alluvium, ascends and crosses a convex surface, and descends beneath side-valley alluvium on the other side of the interfluve (Ruhe, 1959). Compare this geometry with the surface on rhyolite in New Mexico (fig. 7.2).

Sediments above the stone line are sorted, but systematic sampling and laboratory analysis are required to demonstrate this fact. Along the longitudinal profile of the stone-line surface in Iowa, surficial sediments are progressively more clayey downslope, and soils formed in the sediments are in catenary sequence (Ruhe, 1960).

In the southeastern United States sediments above the stone line have been explained by the incorporation of coarse fragments in the subsurface by soil creep. These fragments are detached from dikes or other resistant layers [fig. 7.3(a)] and are drawn along the base of a creeping soil (Sharpe, 1938; Ireland, Sharpe, and Eargle, 1939). Since mass-movement processes cause mixing of particles of all sizes sorting of the material above the stone line cannot be explained by soil creep. Mass movement cannot explain the occurrence of rounded gravel in the stone line where angular gravel particles are supplied from underlying dikes or other resistant layers. It also cannot explain the occurrence of erratic mineral or rock material in the stone line at considerable

Figure 7.4 (a) A diagram showing the Davisian geographical cycle. From W. M. Davis (1899a) **(b) The backwearing of hillslopes. (c) The downwearing of hill summits and hillslopes.** From Holmes (1955), by permission of Am. J. Sci.

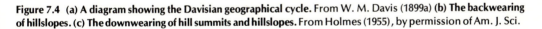

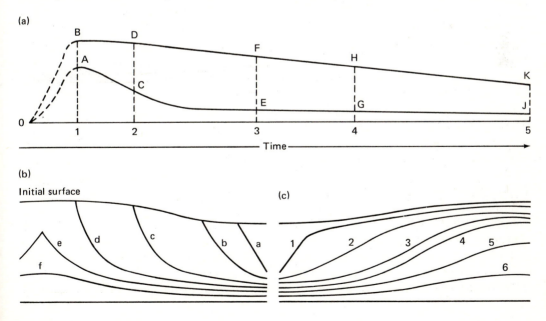

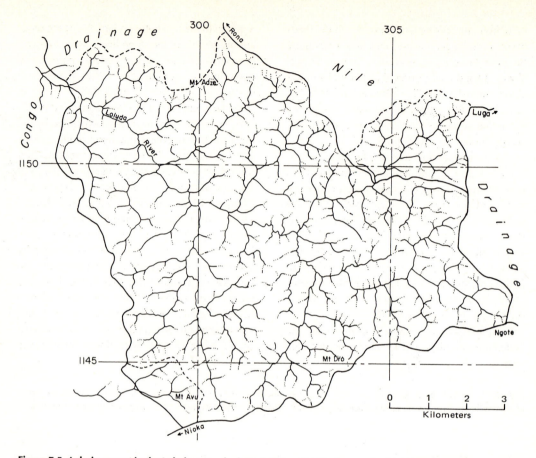

Figure 7.5 A drainage net in the Loluda watershed, Ituri, Zaire. Grid in thousands of meters where 30°E = 200 and equator = 900.

distances from in situ locations and along low gradients. The stone line is an erosion pavement formed by running water on the land surface and concurrently covered by surficial sediment (Parizek and Woodruff, 1957; Ruhe, 1959; Fölster, 1969; Rohdenburg, 1969).

The stone-line surface is similar to that of rock and other pediments and is a pediment itself. The surficial sediment above the stone line is termed pedisediment, indicating its morphostratigraphic relation to the pediment (Ruhe, 1956a; Fölster, 1969; Rohdenburg, 1969).

Kinds and origins of erosion surfaces

A common practice in geomorphology is genetic classification of features which cannot be so named unless they passed through a specific course of processes. Among erosion surfaces, a peneplain is a peneplain because it passed through peneplanation. A pediment is a pediment because it suffered pedimentation. A pediplain is a pediplain because it endured pediplanation.

Generally specific kinds of erosion surface are

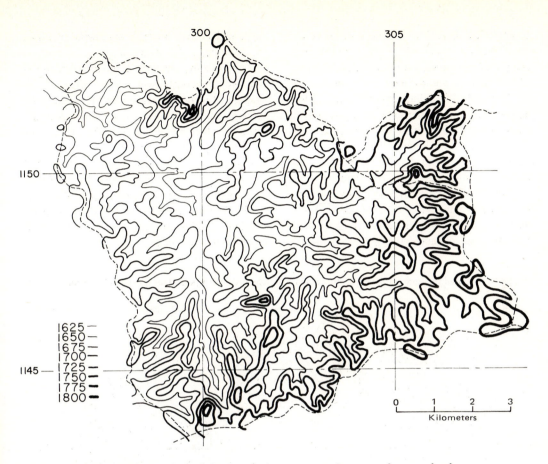

Figure 7.6 An altimetric map of the Loluda watershed. Contours are in meters above sea level.

restricted to specific climatic zones. Peneplains form in humid regions, pediments and pediplains in arid regions. The first requires a humid erosion cycle, the others, an arid erosion cycle (Thornbury, 1969). It is almost heresy to locate a pediment in humid country or a peneplain in arid country.

Peneplains

Peneplain was introduced by William Morris Davis in 1899 as a name for a gently undulating surface of low relief formed by processes of subaerial erosion at the end stage of an erosion cycle. The surface must progress through the Davisian scheme of evolution, which may be visualized diagrammatically [fig. 7.4(a)]. Let a base line from 0 to 5 represent time. At epoch 1 a region of certain structure and form is uplifted, so that *B* represents the average elevation of its higher parts and *A*, the average elevation of its lower parts. The initial average relief is *AB*. Streams rapidly deepen their valleys, and at epoch 2 they have incised to an average elevation represented by *C*. The higher parts of the interstream uplands at epoch 2 have been lowered more slowly to *D*. Relief increases from *AB* to *CD*. Streams then deepen their channels more slowly, as shown by *CEGJ*. Uplands are reduced more rapidly, as shown by *DFHK*. During epochs 1 and 2 valley deepening is most rapid, and during epochs 3 and 4 uplands are reduced most rapidly. During epochs 2 and 3 maximum relief is reached, but during epochs 3 and 4 relief decreases more rapidly than at any other time. After epoch 4 relief is gradually reduced so that only a rolling lowland remains at the end stage of epoch 5. This is the Davisian erosion cycle (Davis, 1899a), and in it are the stages of youth (epochs 1 and 2), matur-

ity (epochs 2 to 4), and old age (beyond epoch 4), culminating in the peneplain.

Note that the lowering of uplands is required contemporaneously with stream incision, although the rates may differ. Consequently, peneplanation is known as the "downwearing" scheme, which may be illustrated by a stream valley and adjacent hill [fig. 7.4(b)]. Note the contemporaneous lowering of the channel, hill summit, and intervening hillslope from an initial surface through phases 1 to 6. This is the "normal" Davisian cycle in the humid environment (Holmes, 1955). Compare the normal soil (Chapter 6) to this scheme.

Figure 7.7 Stepped topographic levels in the Loluda watershed, where a peneplain presumably existed.

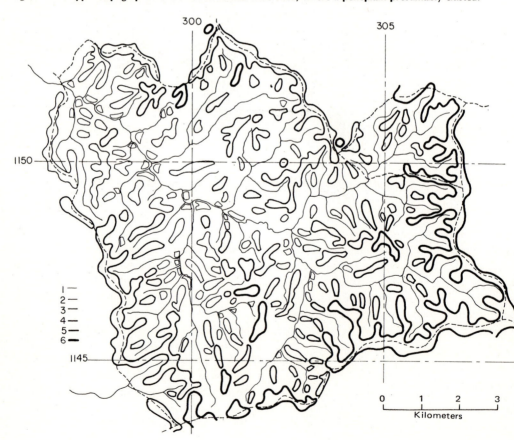

Is the summit lowered contemporaneously with stream incision? An interfluve summit above a side slope or a divide summit above a head slope (fig. 6.2) conforms to the Horton belt of no erosion (fig. 6.7) and has little if any slope gradient. Yet the Zingg, Musgrave, and soil-loss equations demonstrate dependence of erosion upon gradient among other things (Chapter 6). If some gradient does exist, slope length is also a factor in erosion, and at the summit apex, length is zero. If gradient and length are zero as factoral multiples, erosion is zero, and the summit should not wear down.

How do watersheds and drainage basins (Chapter 5) fit on the peneplain or vice versa? Peneplains are large land features such as the Piedmont Plateau (Davis, 1899b), which extends from New York to Alabama and is 10 miles wide at its narrowest part, 125 miles at its widest part (Thornbury, 1965). The African continent has three (Wayland, 1934) or five (Dixey, 1948) extensive peneplains, Jurassic-Cretaceous to end-Tertiary in age. Erosion surfaces in Africa (L. C. King, 1951) occupy the entire continent from 5°N latitude to 35°S latitude and from 12°E longitude to more than 40°E longitude.

Examine a peneplain (Lepersonne, 1949, 1956) in relation to watersheds in the Ituri district of Zaire (formerly the Belgian Congo), west of Lake Albert (Ruhe, 1954, 1956a). The entire Loluda River watershed has been considered as an end-Cretaceous peneplain (Lepersonne, 1956). Part of the perimeter of the watershed is the Congo-Nile continental divide, and the watershed extends almost symmetrically from Ngote northwestward to Shari River of the Congo drainage basin (fig. 7.5). The drainage net in the watershed is dendritic with some tendency toward a parallel pattern of tributaries.

Compare the drainage net (fig. 7.5) to the elevation contours of the watershed (fig. 7.6). Heights of land above 1800 m are at the headward divide at Ngote and at isolated summits toward Nioka, Luga, and Rona. Along the south and north divides, elevations decrease westerly to 1725 m near Mt. Adze and Mt. Avu. Along any interfluve between tributaries of the Loluda River, elevations decrease from the watershed divide toward the main stream. The land surface is scoop- or spoon-shaped much like that described for concave pediments in the southwestern United States (Bryan, 1936; Gilluly, 1937).

Along any interfluve from divide to river, the slope is not continuous but is interrupted like the treads and risers of a staircase (fig. 7.7). Height of land is the divide or the sixth tread or level, and there are five levels below it in sequence along the interfluves. Each level is an erosion surface cut below a higher predecessor (Ruhe, 1956a). What one would expect to be a peneplain or one master erosion surface, when examined in detail within the framework of the watershed, is in fact a stepped sequence of erosion levels. It is difficult to prove that any of the levels originated through the Davisian scheme of peneplanation. A cardinal rule in working with erosion surfaces of any kind is to look within the watershed.

A common feature of the multileveled African surfaces is stone lines with overlying pedisediment (Ruhe, 1959; Collinet, 1969; Lévêque, 1969; Riquier, 1969; Segalen, 1969; Fölster, 1969; Rohdenburg, 1969). In these tropical areas these sediments are deeply and intensely weathered.

Pediments and pediplains

The term *pediment* was first applied to planed bedrock surfaces whose formation was attributed to sheetflood at the foot of mountains in the Sonoran Desert in Arizona (McGee, 1897). Pediments are generally considered landscapes of arid and semiarid country. A commonly used definition states that pediments are erosion surfaces of low relief, partly covered by a veneer of alluvium, that slope away from the base of mountain masses or escarpments in arid and semiarid environments (Hadley, 1967). These requirements are restric-

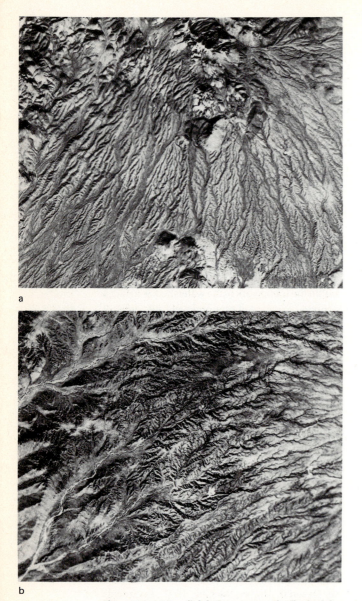

a

b

Figure 7.8 (a) A radial drainage net on the flank of the Dona Ana dome, New Mexico. (b) A dendritic drainage net bordering the Rio Grande in southern New Mexico. Note the attack of side slopes by low-order drainageways and the lateral planation along the major drainageways in both cases.

tive. In desert areas many erosion surfaces are cut in rock and unconsolidated sediments but mountains or escarpments are not involved (fig. 7.2). In moister climate in the Great Plains, pediments commonly flank valleys (Frye, 1954). They are land features in subhumid and humid environments (L. C. King, 1957; Ruhe, 1956b; Hack, 1960; Quinn, 1965; Denny, 1967). To avoid the restrictive definition, a pediment should be considered an erosion surface that lies at the foot of a receded slope, with underlying rocks or sediments that also underlie the upland, which is barren or mantled by alluvial sediment, and which normally has a concave upward longitudinal profile (Howard, 1942).

The planimetric shape of a pediment depends on the pattern of the related drainage net. Where pediments are fan-shaped (Rich, 1935) or narrow in a mountainward direction (Gilluly, 1937), the drainage net is a radial pattern of a distributary system debouching from a mountain canyon into a basin and is similar to the pattern of an alluvial fan. On the pediment the interfluves between the radial drainage lines are erosion surfaces.

Scoop-shaped pediments (Bryan, 1936) conform to dendritic drainage patterns. Lower-order drainage lines join higher-order ones, and one master drainage line controls the entire watershed. The pediment occupies interfluves of the pattern (Gilluly, 1937).

Gross transverse profiles of pediments relate to planimetric patterns. Fan-shaped surfaces have convex cross profiles (Rich, 1935; Howard, 1942), but scoop-shaped surfaces have concave cross profiles (Bryan, 1936; Gilluly, 1937). The usual block diagram of a pediment fronting a mountain range is misleading (Hadley, 1967; Thornbury, 1969). The pediment is usually sketched as a flat surface sloping gently away from a slightly digitated linear mountain front.

A more detailed analysis of cross profiles is required. Examine a radial drainage pattern on a fan-shaped surface [fig. 7.8(a)]. The major drainage lines have smaller tributaries and innumerable low-order drainageways that ascend the side slopes. Three orders of curvature are involved, the first order being the gross convex cross profile. Superimposed is a second-order curvature extending from one major drainageway to an intervening divide to the next major drainageway. A third-order curvature, superimposed on the second-order curvature, consists of minor crenulations from lower-order drainageways to intervening divides. This relation of curved forms to stream orders is important in explaining the pedimentation process.

Examine a scoop-shaped pediment relative to a drainage net [fig. 7.8(b)]. In this dendritic pattern note the major drainageways with smaller tributaries and innumerable low-order drainageways ascending the side slopes. This convergent hydrologic pattern contrasts with the divergent hydrologic fan pattern. Note the sameness of the side-slope patterns in both convergent and divergent systems. In any kind of drainage pattern the pedimentation process must involve the reduction of hillslopes.

Pedimentation is usually explained by two processes or a combination of them: (1) lateral planation by streams, and (2) weathering and removal of debris by rill wash and unconcentrated flow (Hadley, 1967). Inherent in the second explanation is the retreat of slopes (Howard, 1942; Schumm, 1956a, 1962), commonly referred to as the "backwearing" scheme developed by Penck in his *Morphological Analysis of Landforms* (1924). A basal slope forms at the expense of a steeper higher slope that retreats backward. The gentler slope, which replaces the steeper one, develops at the foot of the latter, and flattening of the land progresses during the exchange of slopes. Penckian backwearing opposes downwearing [fig. 7.4(b), (c)].

Backwearing is a part of pedimentation. The steeper part of a hillslope or backslope (figs. 6.2, 6.3) when attacked by rilling and sheetwash retreats into the upland. The basal slope, or footslope, formed at the expense of the retreating

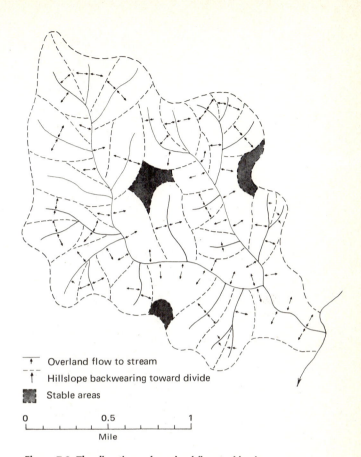

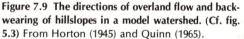

Figure 7.9 The directions of overland flow and backwearing of hillslopes in a model watershed. (Cf. fig. 5.3) From Horton (1945) and Quinn (1965).

backslope, becomes the pediment. The upland, shoulder, and summit (fig. 6.3) are not very eroded but continue to weather. The upland is not lowered until opposing and retreating backslopes intercept one another. The lowered landscape *(f)* [fig. 7.4(c)] is a *pediplain,* which is formed by the coalescence of pediments.

Major work is accomplished by a multidirectional attack on slopes by low-order drainageways accompanied by related hillslope processes (Chapter 6). Within a watershed, overland flow descends from divides between various stream orders to respective members of the drainage net (Horton, 1945). Headward growth of drain-

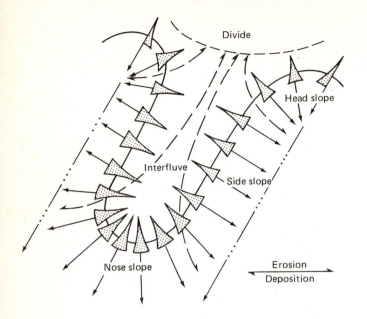

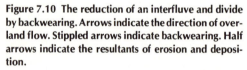

Figure 7.10 The reduction of an interfluve and divide by backwearing. Arrows indicate the direction of overland flow. Stippled arrows indicate backwearing. Half arrows indicate the resultants of erosion and deposition.

ageways proceeds in the opposite direction (Schumm, 1956a), and backwearing of hillslopes proceeds away from the drainageways (Quinn, 1965).

Using the watershed model of stream orders (fig. 5.3), overland flow descends side slopes at right angles to lower drainageways, but backwearing is in the opposite direction (fig. 7.9). Note the numerous directions of flow and retreat, even when restricted to side slopes. The multidirectional attack on interfluves toward divides is more complex when examined in the entire framework of hillslope components (fig. 7.10). Backwearing proceeds from two opposing side slopes and the nose slope, and this concentration lowers the interfluve. The divide, protected between side streams by the interfluve, wears back along the head slope, and a typical digitate upland divide results from localized erosion. The

divide only lowers when hillslopes on the other side of the divide wear back and intercept opposing hillslopes.

In pedimentation lateral planation is probably important in sweeping debris from the base of backwearing hillslopes or in undercutting a hillslope and initiating the backwearing process. Where lateral planation is carried to an extreme, causing coalescence of one planed surface with another, the erosion surface is termed a *panplane* (Crickmay, 1933).

Deep weathering followed by the stripping to lower bedrock is suggested as a process of pedimentation (Mabbutt, 1966). This weathering-stripping mechanism is like one proposed for formation of *etchplains* in Africa (Wayland, 1934). Deep weathering may prepare a landscape for pedimentation in humid areas but hardly in arid and semiarid environments, where weathering is usually shallow.

Stepped erosion surfaces

It is rare for a single erosion surface to occur in an area or region. Usually multiple surfaces are arranged in steps on the landscape, and they occur worldwide (Geyl, 1961). The Loluda Watershed in Africa exemplifies stepped erosion surfaces in a humid area, but they are common in arid country as well [fig. 7.2(b)]. To delineate stepped surfaces, look within the watershed regardless of its size.

Along the Rio Grande in southern New Mexico, arroyos descend alluvial fans from bordering mountains to the valley (fig. 7.11). On interfluves between arroyos and from 3900 to 4300 feet in elevation, there are five surfaces in stepped sequence away from the valley. Each level has been recognized as a *formal geomorphic surface* (Ruhe, 1967), which is a part of the land that is specifically defined in space and time and may include many landforms. It is a mappable feature whose geographic distribution is portrayed on maps or air photographs and whose geometric

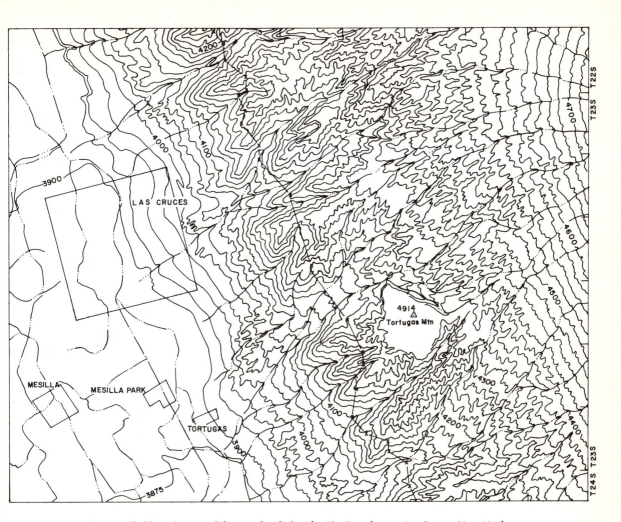

Figure 7.11 A drainage and altimetric map of the area bordering the Rio Grande near Las Cruces, New Mexico. Scale: 1 mile between ticks on map right margin.

dimensions are specified and analyzed. Its association with other geomorphic surfaces is defined in order to place it in a space and time sequence. Its association with the rock or sediments below it or on it is specified. It is dated by relative or absolute means, and the dating is defined. The geomorphic surface is labeled with a geographic name (Ruhe, 1969a).

The surfaces in New Mexico in order upward are Fillmore, Leasburg, Picacho, Tortugas, and Jornada (fig. 7.12). Each lower level is cut into and below its predecessor, and each surface has a distinct backslope and footslope. Although cut in unconsolidated sands and gravels, these erosion surfaces are pediments in this arid country where present annual rainfall is about 7 inches.

Look at a watershed in Iowa where current annual rainfall is about 32 inches. The landscape is

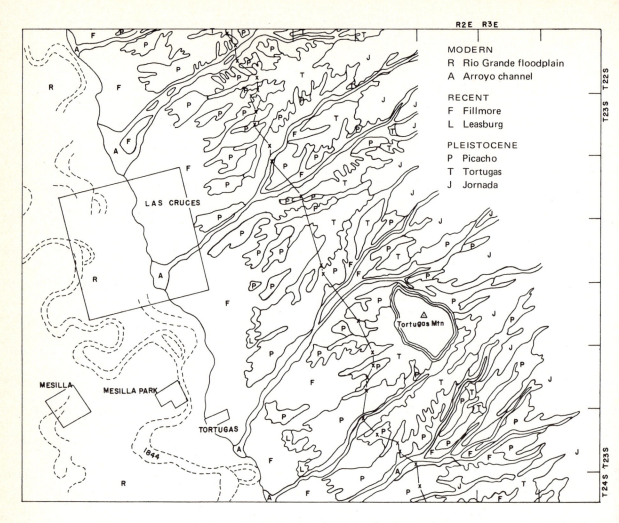

Figure 7.12 Erosional and constructional surfaces bordering the Rio Grande near Las Cruces, New Mexico. From Ruhe (1967), by permission of N. Mex. Bur. Mines and Miner. Resour.

stepped along interfluves. There are three levels from A to B but only two levels from C to D [fig. 7.13(b)]. The uppermost surface, buried beneath loess, is the relict Kansan drift plain with a well-developed Yarmouth-Sangamon paleosol on it. Each lower level is cut in Kansan till below its predecessor and has a concave backslope rising to the next higher level and a descending footslope (fig. 7.14). The soils on each surface are distinctly different (Ruhe, 1956b).

Examine an intermountain basin between the Koolau and Waianae Ranges on the island of Oahu, Hawaii, where annual rainfall ranges from 30 to 100 inches (fig. 7.15). Watersheds descend eastward from the Waianae Mountains and westward from the Koolau Mountains and then southward to Pearl Harbor. The Waikele, Kipapa, Poliwai, and Waiawa gulches are incised deeply, and tributary drainageways hang at the gulch edges. Between major drainageways interfluves ascend from a scarp at 100 feet above sea level to 1200 feet (fig. 7.16) and have six stepped levels toward the Waianae Range and eight stepped levels toward the Koolau Range (fig. 7.17). Each lower surface is cut below its predecessor [fig. 7.1(c)]. The Kipapa, Waiahole, Schofield, and Waikakalaua surfaces are cut in basalt flows that emanated from the Koolau Range and a conglomerate built as an alluvial fan from the Waianae Range. This lack of discrimination is prime evidence of erosion surfaces.

Area-altitude analysis shows a rise of these levels up the interfluves. Measure the areas by increments of elevation and plot as a frequency distribution for each surface (fig. 7.18). Mean elevations differ from the lowest to the highest surface, but standard deviations and ranges progressively overlap. This reduction of three to two dimensions simplifies visualization of relations among all of the surfaces.

There are similar stepped erosion surfaces in different climatic and vegetation zones from arid desert shrub to subhumid prairie and humid tropical savanna. The surfaces are cut in materials

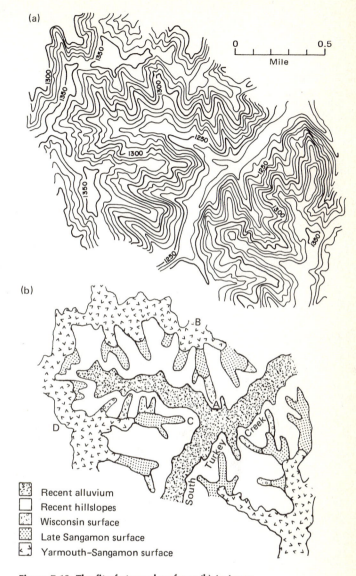

(a)

(b)

Recent alluvium
Recent hillslopes
Wisconsin surface
Late Sangamon surface
Yarmouth–Sangamon surface

Figure 7.13 The fit of stepped surfaces (b) to topography (a) in the Adair quadrangle, Iowa. From Ruhe (1969a) by permission. © Iowa State University Press, Ames.

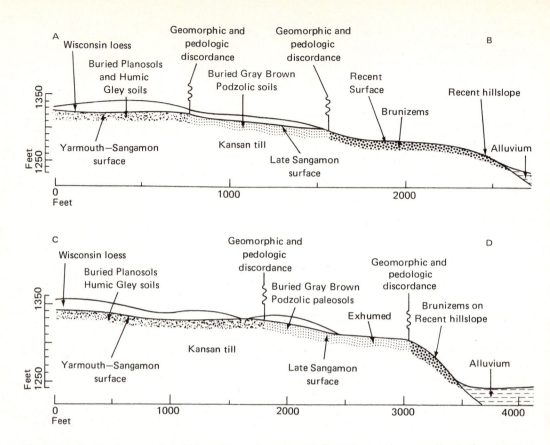

Figure 7.14 Profiles from C to D and A to B of the surfaces shown in fig. 7.13. Note the stepped levels, the curvatures of the surfaces, and the complex soil relations. From Ruhe (1969a), by permission. © Iowa State University Press, Ames.

ranging from unconsolidated sands, gravels, and glacial till to igneous and metamorphic rock. The surfaces have similar form and spatial arrangement in watersheds, and a certain universality exists among them regardless of climate or material (L. C. King, 1957). Is it still necessary to discriminate between "arid" and "humid" geomorphology and not to recognize pedimentation in all climatic zones?

Stepped topography on granitic rocks in the Sierra Nevada has been explained by differential weathering (Wahrhaftig, 1965). Weathered debris is eroded, which leaves a barren rock surface at a higher level. This process is a variant of the

etchplain scheme (Wayland, 1934) and the mantle control of pedimentation (Mabbutt, 1966). It is difficult to explain the spatial arrangement of erosion surfaces within watersheds, erosion pavements, sorted surficial sediments, and hydraulic-type profiles by any etching process.

Stepped erosion surfaces demonstrate an age sequence, *if* major complications can be negated. The surfaces must not be controlled by rock or other geologic structure. If a surface coincides with a resistant rock stratum, it is a *stripped or structural plain*. If the stepped levels are erosion surfaces, the highest level is the oldest, the lowest level the youngest.

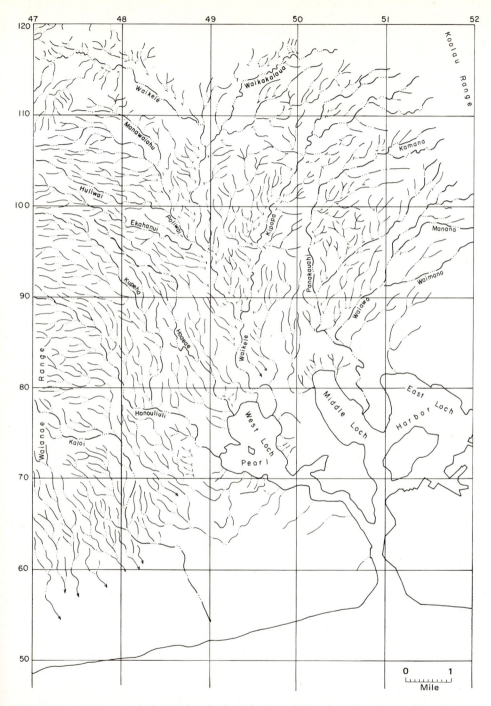

Figure 7.15 A drainage net in the Wahiawa basin, Oahu, Hawaii. Hanging valleys along gulch walls are shown by the bar terminating the tributary.

Soils on erosion surfaces

Recall the Jenny soil function (Chapter 2) and fit it to erosion surfaces in stepped sequence that range through time. The time factor varies and affects soils on different surfaces. Climate and associated vegetation may change through time, and topography may also. Although the parent material is least susceptible to change through time, other factors of the soil function do change, and soils may consequently differ from surface to surface.

Lithologic discontinuity

The parent material can differ on an erosion surface where pedisediment overlying a stone line thickens downslope. Where pedisediment is thin, part of the solum may form in it, and another part may form in material beneath the stone line. Downslope, where pedisediment thickens, the entire solum may form in the sediment. Between these two sites, an increasing number of soil horizons may form in pedisediment.

This difference in soil morphology is recognized in soil-horizon nomenclature. If an *A* horizon and the upper part of a *B* horizon are formed in pedisediment and the lower part of the *B* horizon and the *C* horizon are in substrata beneath a stone line, the soil profile has *A1, A3, B1, IIB2, IIB3,* and *IIC* horizons. The numeral *II*, prefixed to all soil horizons beneath the stone line, shows that a soil has more than one parent material. The materials are interrupted by *lithologic discontinuity,* which is marked by the stone line.

Multilayered parent material on an erosion surface can differ physically, chemically, and mineralogically. On the pediments in the Loluda watershed in Africa, the bedrock in places is weathered granite or mica schist. The pedisediment above the stone line is derived from laterite crusts cut by the pediment backslope. Soils formed in this previously thoroughly weathered lateritic debris have properties very different from those of soils formed from granite and mica schist (Ruhe, 1956a).

Differential sorting of pedisediment on an erosion surface may cause changes in texture, consistence, bulk density, porosity, and permeability at the lithologic discontinuity, which can affect formation of soil horizons. On the pediments in Iowa, there is no mineralogic difference between pedisediment and the lower Kansan till, but the physical differences locally control eluviation and illuviation (p. 38) so that the soil horizon just beneath the lithologic discontinuity has maximum clay accumulation (Ruhe, 1960).

Geomorphic and time relations

In soil-geomorphology research two independent mapping programs are conducted. In one the geomorphic surfaces are mapped, and in the other the soils are mapped. Comparison is made without one mapping biasing the other. Usually there is a fit of soil mapping units to geomorphic surfaces, but the fit may not be exact. Compare the soil association areas (fig. 7.19) with the geomorphic surfaces in Oahu, Hawaii (fig. 7.17). The stepped surfaces, from Mahoe to Kamana, rise from Pearl Harbor toward the Waianae and Koolau Ranges in a concentric, arcuate pattern. The soil association areas, each named for a major soil, have a similar banded concentric pattern.

A first correlation is made. Kunia and Mahana soils are on surfaces in an alluvial-fan conglomerate that formed east of the Waianae Range. Other soils—Molokai, Lahaina, Wahiawa, Leilehua, Waipio, Manana, and Paaloa—formed on surfaces in Koolau basalts. Difference in parent material is the basis for this broad separation.

Note that in the Kunia soil-association area (fig. 7.19) the erosion levels Pohaku through Schofield are along some interfluves (fig. 7.17). Along the interfluve between the Waikele and Kipapa gulches (fig. 7.15), an extensive area of

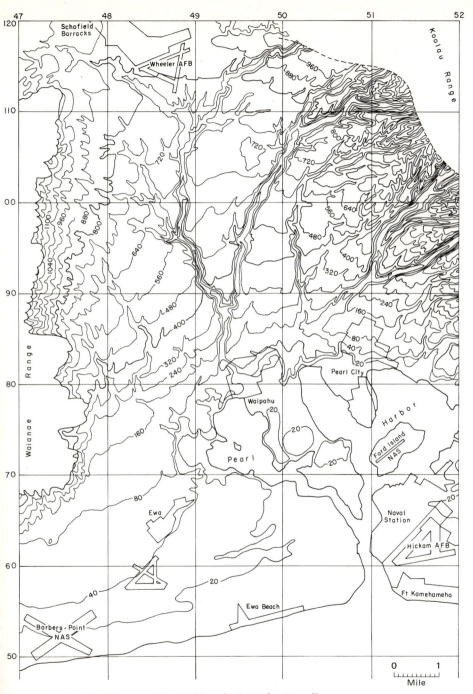

Figure 7.16 An altimetric map of the Wahiawa basin, Oahu, Hawaii.

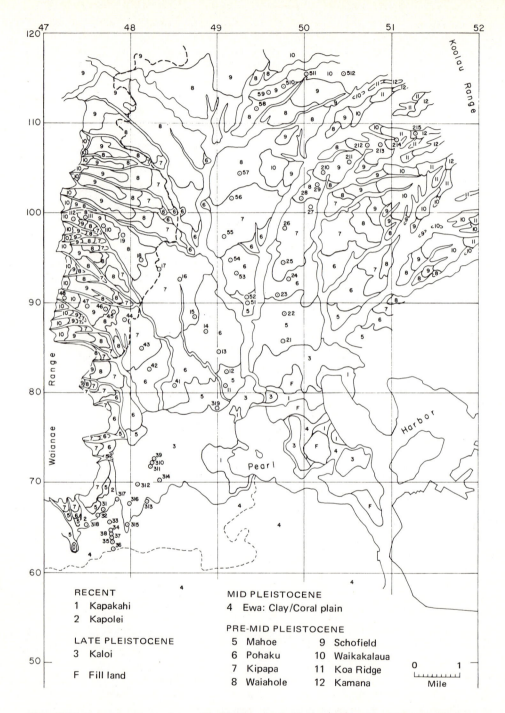

Figure 7.17 Stepped erosion surfaces of the Wahiawa basin. The broken line shows the contact between the fanglomerate from the Waianae Range and the interfingered basalt flows from the Koolau Range. Circles indicate borehole locations.

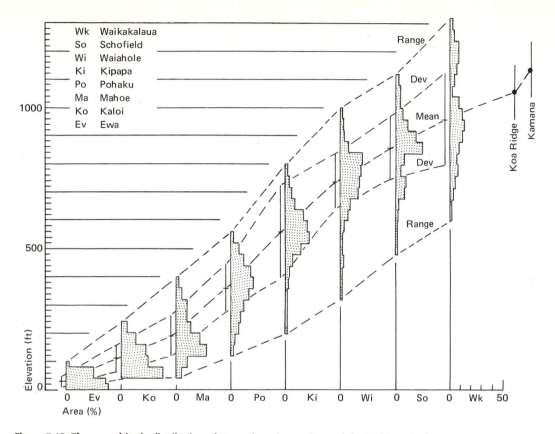

Figure 7.18 The area-altitude distribution of stepped erosion surfaces of the Wahiawa basin.

Wahiawa soils has stepped levels from Kipapa through Schofield. Any soil-association area crosses several levels of the stepped sequence, but a dominant part of any soil association relates to a specific erosion level (table 7.1). The frequency mode of Molokai soils correlates with the Pohaku surface, but these soils are also on the younger Mahoe and the older Kipapa surfaces. Lahaina soils dominate the Kipapa surface, but they also occur on the younger Pohaku and older Waiahole surfaces, and so on.

The frequency mode of soils (table 7.1) also correlates with increasing mean elevation of erosion surfaces (fig. 7.18). Soil-association areas are concentrically banded with elevation, which in turn creates a relation to climatic zones. Current

rainfall patterns are concentrically banded in the intermountain basin, with rainfall increasing from lower to higher elevation. Differences in soils are due to differences in climate (Cline, 1955).

Can the time factor from younger to older stepped surfaces be excluded entirely? Mahoe, the lowest and youngest surface, stands above the Kaena shoreline, 95 feet above sea level, of probable mid-Pleistocene age (Stearns, 1961). The soils on all the other surfaces are therefore relatively old and are extremely weathered, being essentially sesquioxides and clay minerals (Juang and Uehara, 1968). But climatic zoning in the past cannot be excluded, when different climates related to different stands of sea level (Ruhe, 1964

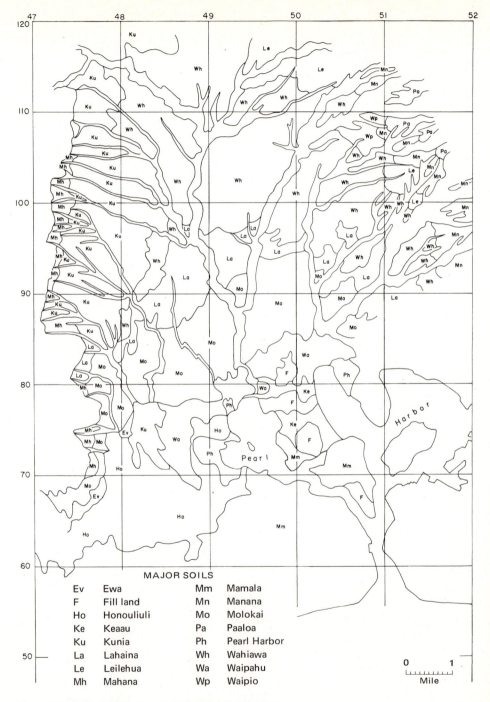

Figure 7.19 Soil-association areas in the Wahiawa basin.

MAJOR SOILS

Ev	Ewa	Mm	Mamala
F	Fill land	Mn	Manana
Ho	Honouliuli	Mo	Molokai
Ke	Keaau	Pa	Paaloa
Ku	Kunia	Ph	Pearl Harbor
La	Lahaina	Wh	Wahiawa
Le	Leilehua	Wa	Waipahu
Mh	Mahana	Wp	Waipio

0   1
Mile

Table 7.1 Areal distribution of soils on geomorphic surfaces, Oahu, Hawaii

Soil	Total area (acres)	Geomorphic surface*(%)							
		5	6	7	8	9	10	11	12
Molokai	4730	38.0	57.6	4.4					
Lahaina	3082		39.8	50.6	9.7				
Wahiawa	8705		2.3	26.8	46.7	18.5	5.8		
Leilehua	1062				6.0	29.6	64.4		
Waipio	200						85.7	14.3	
Kunia	3750		1.1	21.4	32.4	45.0			
Mahana	816					15.7	84.3		
Manana	597						41.7	56.4	1.9
Paaloa	75							25.0	75.0

*The surfaces are Mahoe (5), Pohaku (6), Kipapa (7), Waiahole (8), Schofield (9), Waikakalaua (10), Koa Ridge (11), and Kamana (12).

Table 7.2 Comparison of properties of soils on stepped surfaces in Iowa

Surface	Soil	Thickness of solum (in.)	Thickness of B horizon (in.)	Clay in B horizon (%)	Soil horizon*	Wrh index[†]	Wrl index[††]
Recent	A	15	11	31.2	A	0.79	2.09
	B	32	23	32.2	B	0.92	2.13
	C	29	22	34.6	C	0.68	2.21
	Average	25	19	32.3			
Late Sangamon	D	46	32	50.7	A	1.27	3.06
	E	70	56	49.1	B	1.12	2.49
	F	39	29	49.5	C	0.77	2.04
	Average	52	39	49.7			
Yarmouth-Sangamon	G	87	70	51.4	A	2.11	4.85
	H	68	44	57.7	B	1.62	3.00
	I	85	62	50.7	C	1.28	2.57
	Average	80	59	53.2			

From Ruhe (1956b), by permission. © The Williams & Wilkins Co., Baltimore.
*Averages of A, B, and C horizons of soil groups A-B-C, D-E-F, and G-H-I.
[†] Wrh = zircon + tourmaline/(amphiboles + pyroxene).
[††] Wrl = quartz/feldspars.

b, 1965b): This Hawaiian study illustrates the difficulty of filtering factors of soil formation in functional analysis. The surfaces represent a time sequence, but environment and other factors also changed through time.

Stepped surfaces in Iowa are more diversely separated in time, the oldest surface being as old as mid-Pleistocene (fig. 7.14) and the youngest surface being Recent. The intermediate surface is late Pleistocene in age, and all surfaces are cut in a common parent material, Kansan till.

Soils differ distinctly (table 7.2), and as surfaces become progressively older, the thickness of the solum increases, the thickness of the B horizon increases, the clay content of the B horizon increases, and heavy and light mineral weathering indices increase (Chapter 2). Environmental factors are not the same. Soils on the Recent surface formed under grass. Soils on the late Pleistocene surface formed under trees. Soils on the mid-Pleistocene surface could have formed under grass or trees. A forest-soil analogue on the Recent surface has a development comparable to the soil formed under grass (Prill and Riecken, 1958). Soil differences on the stepped surfaces in Iowa thus depend mainly on time or duration of weathering.

Soil development on stepped surfaces in New Mexico relates to increasing age, as shown by increasing carbonate accumulation, silicate clay accumulation in B horizons, and thickness of the solum (Gile, 1970). The Fillmore surface (fig. 7.12) dates from a charcoal hearth 46 inches beneath the surface at 2620 ± 200 years. The Picacho surface is older than 9550 ± 300 years, as dated from soil organic carbon in a B horizon that was later cemented by carbonate, forming calcrete (Ruhe, 1967). Soils on the Fillmore surface have little clay accumulation and no textural B horizons. Carbonate is sparse, and original C horizon fabric is present. Soils on the Picacho surface have prominant textural B horizons and strong carbonate horizons (Gile, 1970) that are very distinct (fig. 2.13).

Geomorphic surfaces provide a chronological framework in soil studies if other factors are filtered from the analytical system. A given surface may have many kinds of soils, but they will have a common degree of soil development. Soils usually display increasing development from the youngest to the oldest surface. The following corollary may be drawn: after proceeding through the analysis of surface to soil, soil may be used to delineate surface.

Several of the major agents responsible for geomorphic processes are independent of physiographic control. Wind, for example, is independent of any topographic divide and can blow across it. Consequently, its effects can be widespread, and the natural boundaries of eolian landscapes are sometimes difficult to define.

Aerodynamics of wind and particles

Wind is air in motion and, like fluid in motion, has the ability to do work, including erosion, sediment transport, and deposition. Velocity is a controlling factor in the aerodynamic system and was thoroughly investigated in a classic study on blown sand and desert dunes (Bagnold, 1941). Liberal use will be made of this study, as well as the many basic principles of fluid dynamics that apply to aerodynamics, in the following sections.

Wind velocity near ground

Air flow, like fluid flow, may be laminar or turbulent, but any near-surface flow is turbulent because of ground roughness. Near-surface wind velocity is proportional to the logarithm of height above ground, and a vertical velocity curve is similar to one beneath a free-water surface (fig. 3.3). The formula for velocity (V) at height z above ground is:

$$V = 5.75 \, Vd \, \log \frac{z}{k}$$

8

Wind and Eolian Landscapes

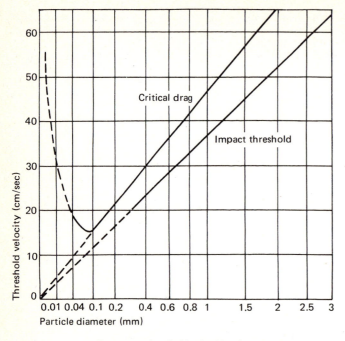

Figure 8.1 Threshold velocities that cause movement of particles. Particle size is plotted as square-root scale. From Bagnold (1941).

where Vd is drag velocity and k is a coefficient of surface roughness. The formula for drag velocity is:

$$Vd = \sqrt{\frac{Bd}{da}}$$

where Bd is bottom drag force and da is density of air.

If particles are moved in flowing air, they affect flow, and velocity is altered. Since velocity about 3 mm above the ground remains almost constant no matter how hard the wind blows, the velocity equation is:

$$V = 5.75 \, Vd' \log \frac{z}{k'} + Vt$$

where Vt is the threshold velocity at the critical height k' of 3 mm. Velocity Vt is caused by impact of one grain on another.

Like a particle protruding above the mean bed level in a stream and acted upon by flowing water (fig. 3.8), particles protruding above the mean ground level are acted upon by wind. The threshold velocity for critical drag on a particle is expressed by the equation:

$$Vd = A \sqrt{\frac{dp - da}{da} \, gD}$$

where dp is the density of the particle; da, the density of the air; g, acceleration of gravity; and D, the particle diameter. The constant A is about 0.1 for air, and if the value of A remains the same, the threshold value of Vd varies as the square root of the particle diameter.

If critical drag is substituted in the original velocity equation, threshold velocity, Vt, is obtained:

$$Vt = 5.75 \, A \sqrt{\frac{dp - da}{da} \, gD \log \frac{z}{k}}$$

When critical drag and impact threshold velocities are plotted against the square root of the particle size (fig. 8.1), the importance of velocity in movement and transport of particles is seen. Note the similarity to the Hjulström diagram (fig. 3.11).

Just as velocity is important in fluid motion, the rate of movement of air is critical in eolian erosion and transport of sediment. After particles are entrained at a given velocity, they are maintained in transport at lower velocities, but deposition occurs with a further reduction or check in velocity.

Eolian erosion, sediment transport, and deposition

Wind erosion occurs at the instant that a grain is moved. Movement begins when the critical drag velocity is reached for a given-sized particle (fig. 8.1). Some small grains become airborne, remain in *suspension* in the atmosphere, and are carried downwind for considerable distances. Some larger grains become airborne but settle to earth slightly downwind. These grains move in a series of bounds, or *saltation*. Upon returning to the ground they may strike other particles whose impact threshold velocities are reached, which causes them to become airborne either in suspension or saltation. But impact threshold velocity may not be reached by some particles, and they slide or roll along the surface in *reptation* or *surface creep*. About 25 percent of dune sand movement is by creep, with the remainder by saltation (Bagnold, 1941).

During saltation, paths of bounding grains vary in height and length of trajectory, but the angle of ground impact is almost constant. When a grain reaches the apex of a trajectory, it tends to fall, following Stokes' law. When the resistance to settling and the net settling force are equal, the terminal velocity becomes constant and acts vertically downward. Wind acts on the grain as a vector parallel to the ground, so the grain must strike the ground at an impact angle along the resultant of the terminal velocity and wind velocity vectors. The impact angle varies from 10° to 16° for a wide range of conditions (Bagnold, 1941).

Wind can move particles larger than sand along the ground if surface conditions are favorable. Wind speeds of 60 to 75 km/hr have been known to move pebbles weighing 4 to 56 g up an inclined glazed surface (table 8.1). This is an unusual example of creep transport.

The bulk of the saltation load is concentrated within 1 m above ground, but this layer may be stratified. Examine data from the desert of southern New Mexico, where dust traps 2 ft × 2 ft × 4 in. deep were placed 1 foot and 3 feet above ground (table 8.2). These open pans were filled with plastic spheres, and trapped particles fell vertically or along saltation trajectories. Note that sand was trapped at each height but was more abundant in the lower trap. Dust (silt and clay) was more abundant in the upper trap.

Table 8.1 Size and weight of particles moved by wind on a glazed surface

Particle	Size (cm)	Weight (g)	Wind velocity (km/hr)		
			60	70	75
				(inclined surface°)	
A	4.9 × 3.9 × 2.7	56	—	1	5
B	3.8 × 2.9 × 2.0	33	1	5	7
C	3.0 × 2.5 × 1.9	24	1	6	9
D	2.6 × 2.0 × 1.6	12	3	16	19
E	1.8 × 1.8 × 0.9	4	5	20	20

From Schumm (1956b), by permission of Soc. Econ. Paleontol. and Mineral.

Wind erosion and transport sort material, and the modal diameter of the frequency distribution of size is a key to the sorting process (Chepil, 1959). Sizes larger than the modal diameter of saltating grains lag, and those smaller deflate and are carried for a distance. Depending on the texture class, about 31 to 78 percent of particles less than 0.1 mm deflate in a single windstorm. Silt deflates more readily than sand or clay, and in coarser materials finer particles are removed with sand particles lagging.

Deposition of sand or dust particles is caused by a reduction or check in wind velocity, and three processes are active (Bagnold, 1941). *True sedimentation* occurs when silt and clay particles settle through slowly moving air and when their impact velocity is insufficient to move another particle. *Accretion* occurs when the surface wind velocity decreases or when the surface roughness changes. Saltating grains may cause others to move, but more grains come to rest than are moved. The roughness of surface microdepressions or vegetation provides sites for immobilizing grains. *Encroachment* occurs when obstructions are present or when slope facets change the surface profile. Creep is diminished, but saltation continues. Vegetation halts creep; also grains may roll down a declivity and come to rest, where they are sheltered from impact. Surface wind velocity need not decrease.

Wind creates landforms and landscapes by erosion and deposition, and the most important features are constructed during deposition of wind-borne sediment.

Features of wind erosion

Wind erodes by *deflation,* the removal of loose material from the ground, and by *abrasion,* the impact of particles against objects. When deflation is concentrated locally a depression forms as material is blown away which is called a *blowout* or *deflation basin*.

The effects of wind abrasion are best visualized in the industrial process of sandblasting. Particles are discharged at high velocity and can remove baked molding sand from a steel casting and grime from the stones of a building. Similarly, wind-blown sand abrades the rock that it impacts. Since the bulk of sand is carried by wind near the ground, it undercuts a vertical rock column, forming a *pedestal rock*.

Wind can abrade a flat surface on a cobble projecting above ground level. After partial rotation of the cobble by frost action or some other process, another flat face may be abraded so that the faces intersect at an edge. A faceted cobble is a

ventifact and may have many intersecting faces. If fine-grained, the surfaces may be *polished,* or have high luster. The impact of grains on mineral or rock material with glassy luster causes *frosting*.

These erosive effects are of minor importance, however, when compared to the constructional landforms and landscapes, namely dunes and loess.

Dunes

Wind deposits occupy large regions of the earth's surface and are distinguished primarily by particle size. Hills formed by wind-blown sand are *dunes;* deposits formed mainly by wind-blown silt are *loess.* About 30 percent of the United States is covered by eolian sediments and associated landforms (fig. 8.2).

The building of *active dunes* requires a continuing source of sand, reasonably barren ground, and adequate winds. Vegetation anchors sand in *stabilized dunes*.

Dune morphology

Dunes, which can be simple or complex features, are classified by form (Melton, 1940; Hack,

Table 8.2   Dust collection in traps in a desert in southern New Mexico from February to June*

Trap	Weight	Particle size (mm)					Organic carbon		Carbonates	
		2.0-0.25	0.25-0.1	0.1-0.05	0.05-0.002	less than 0.002				
	(g/m²/yr)	(%)	(%)	(%)	(%)	(%)	(%)	(g/m²/yr)	(%)	(g/m²/yr)
Basin site										
3 ft	20	1	17	9	34	39	4.7	0.8	1.4	0.2
1 ft	42	6	32	12	26	24	3.1	0.9	0.8	0.3
Fan site										
3 ft	29	2	20	22	34	22	3.6	0.9	1.5	0.4
1 ft	164	6	35	29	17	13	1.0	1.4	0.3	0.4

From Gile, Hawley, and Grossman (1970).
*The data for the 3-feet height is for the years 1963-1969; those for the 1-foot height, for 1967-1969.

Figure 8.2 Wind-blown sediments forming surficial deposits in the United States. From Thorp and Smith (1952), by permission of Geol. Soc. Am.

1941). The ground plan of *barchan dunes* is crescent-shaped with a convex trailing edge but with the crescent's horns pointing downwind. They have an asymmetrical cross profile with a gentler upwind slope, which rises to a crest and then descends slightly to the brink of a steeper slope, or *slip face,* descending downwind [fig. 8.3(a)].

Barchans usually form where sand is in transport across barren nonsandy ground (Melton, 1940). Upon deposition of a pile of sand, the form is streamlined in vertical and horizontal planes by wind from a constant direction (Bagnold, 1941). Sand moves by creep and saltation up a windward slope. Eddies form leeward of the crest and steepen the leeward slope to a gradient at which the sand grains fall down the slip face and come

to rest at an angle of repose. As the dune height increases, resistance to the airstream increases, but the horizontal streamlines have less resistance near the periphery of the sand pile, where dune heights decrease to zero. At these extremities, the horns of the crescent form and advance more rapidly than the center of the dune.

Barchans advance by a process of subtraction and addition. Sand is stripped from the windward slope and dumped down the leeward slope (Bagnold, 1941). The rate of advance depends on the direction, velocity, and duration of winds, the supply of sand, the nature of the surface being crossed, vegetation, and the size of the dune. The height of the dune is the most important size factor. The advance of barchans in Peru (Finkel, 1959) and in the Imperial Valley of California

Figure 8.3 Dune forms in southern New Mexico. (a) Barchan dunes at White Sands. (b) Coppice dunes near Organ. (c) Longitudinal dunes near San Diego Mountain. Scale in miles.

(Long and Sharp, 1964) relates to dune height in the equation:

$$\frac{1}{D} = a + bH$$

where D is displacement, or distance, and H is height. In Peru H is measured from dune crest to dune base, in California from the brink of the slip face to the dune base. In California 47 dunes moved 325 to 925 feet between 1956 and 1963, averaging 82 feet per year. Between 1941 and 1956, 34 dunes moved 350 to 1200 feet, averaging 50 feet per year (Long and Sharp, 1964).

Barchan size is measured by height and three map distances: from the trailing edge to the brink of the slip face along an axis of symmetry (a), from a tangent to the trailing edge to the tip of the horns of the crescent (b), and between the tips of the horns (c). In California the advance of slip faces was 2.4 to 12.2 feet per month; H, 31 to 37 feet; a, 179 to 258 feet; and c, 520 to 723 feet (Norris, 1966). Dune volumes varied from 1.2 to 1.6 million cubic feet.

Transverse dunes are ridges of sand whose parallel crests are perpendicular to wind direction and whose profile has a gentle windward slope and a steeper leeward slip face. Parallel ridge crests are usually evenly spaced. These dunes form where winds blow in a constant direction across a large area of barren, loose sand (Melton, 1940).

Longitudinal or *seif dunes* are long parallel ridges aligned with wind direction [fig. 8.3(c)]. Where wind direction is constant and sand is abundant, they form in the wind shadow of an obstacle and streamline downwind (Melton, 1940). They may form from barchans in bidirectional wind (Bagnold, 1941). A quartering wind (a wind that comes in at an angle) causes elongation of one horn, and after the wind shifts to its original direction, the prolonged horn advances farther. The longitudinal axis of the seif dune follows the resultant between the original and quartering wind vectors.

Longitudinal dunes may form by the breeching of the headward closure of U-shaped dunes (H. T. U. Smith, 1965). The bend of the U is removed, leaving two parallel linear ridges.

A *U-shaped dune* is also known as a *parabolic, blowout,* or *upsiloidal dune* (Melton, 1940; H. T. U. Smith, 1941). The closed end points downwind. A U-shaped dune may form from a barchan or any other kind of dune. Deflation on the windward slope forms a blowout depression, and sand is transported across the crest and dumped down the slip face. The lateral concave slip face of a barchan and the linear slip face of a transverse dune change to lateral convex curvature by the downwind additions. These dunes also form by the mergence of two ridges downwind caused by slight changes in wind direction (Melton, 1940; H. T. U. Smith, 1949).

Coppice dunes [fig. 8.3(b)] are mounds of sand associated with bunch or clump vegetation (Melton, 1940). In the southwestern United States, mesquite grows readily on loose sand and is not killed by slow burial. The shrub holds the sand mound together, and large areas of the mounds form a *patterned ground*. In southern New Mexico coppice dunes 1 to 8 feet high and 5 to 40 feet in diameter have formed since 1885 following destruction of desert grassland (Gile, 1966).

The internal structure of a sand dune relates to its external form and the processes of formation. The sorting and grading of particles in thin beds on low-angle windward slopes and steeper-angle slip faces, create graded and cross-bedding [fig. 8.4(a), (b)].

Although siliceous minerals are more common, any sand-size material can form dunes. In Bermuda the islands are essentially dunes formed on paleosols on older dunes [fig. 8.4 (c)] or on marine limestones (Sayles, 1931), and they are composed of fragments of shells and carbonate rock derived from beaches. The dunes were secondarily cemented by carbonate during weathering (Ruhe et al., 1961) to form *eolianite* (Sayles, 1931).

The White Sands in New Mexico are barchans composed of gypsum sand [fig. 8.3(a)]. Calcium sulfate precipitates in the playa lake, Lucero, as an evaporite, and during dry-lake stages sand-size gypsum particles are transported downwind, forming the dunes of White Sands National Monument.

Ripples

The surface of a barren sand dune is usually rippled [fig. 8.3(a)]. The cross section of a ripple is like that of the dune, with a shallower windward slope and a steeper leeward slope. Ripples are spaced with uniform wave length, which depends on wind strength and characteristic grain paths of saltation that cause surface creep (Bagnold, 1941). Ripple height and shape depend on the particle size of the sand (Sharp, 1963). Since the length of bound and the angle of impact of a saltating grain depend on wind velocity and since ripple height depends on grain size, the gradient of the leeward ripple slope equals the incident angle of impact.

Ripples advance across dunes by creep, and displacement is directly related to wind velocity

Figure 8.4 The internal structures of eolianite dunes in Bermuda. (a) Cross-bedded eolianite. (b) An eolianite dune on buried soil (P) with subjacent filled solution pipes in dipping eolianite. (c) Detail of cross-bedding. (d) The banded soil B horizon in a dune in North Carolina.

(Sharp, 1963). On the Kelso Dunes in the Mojave Desert in California, ripples were seen to be stationary at a wind velocity of 15.5 miles per hour measured 4 feet above the dune surface, but at a wind velocity of 30 miles per hour the ripple advance was 2 inches per minute. Since the wind velocity 3 mm above the ground is almost constant no matter how hard the wind blows (Bagnold, 1941), critical drag velocity (fig. 8.1) must

have little effect in forming ripples—impact threshold velocity must be responsible.

Implications of dunes

Dunes show the directions of dune-building winds and may permit interpretation of the climate under which the landscape formed (H. T. U. Smith, 1949). The horns of symmetrical barchans point downwind, and an axis of symmetry bisecting the crescent approximates the wind azimuth. The axis of longitudinal dunes and a perpendicular to the crest line of transverse dunes parallel air flow. The closure of U-shaped dunes points downwind. Since coppice dunes are domed and circular, their shape offers no clue to wind direction. Internally shallow dipping beds point windward, and steeply dipping beds point leeward. In other dune forms the externally shallow slopes face upwind, and the steep slip face is downwind.

In southern New Mexico parallel longitudinal dunes indicate wind direction from southwest to northeast [fig. 8.3(c)]. On the High Plains near Lovington, New Mexico, soil patterns show an ancient longitudinal dune field trending northwest to southeast with corresponding wind direction (Price, 1958). Soils have indurated carbonate horizons, and where upper soil horizons have been stripped, calcrete is at the surface, forming scabland (fig. 8.5). In alternate bands calcrete is at depths of 6 feet in Lea soils and more than 6 feet in the Springer and Reagan soils. In a profile cross-

Figure 8.5 The soil pattern in an ancient longitudinal dune field on the High Plains near Lovington, New Mexico: S, scabland (calcrete at the surface); P and R, Springer and Reagan soils (calcrete at a depth greater than 6 feet); L, Lea soils (calcrete at a depth of about 6 feet). From Harper and Smith (1932).

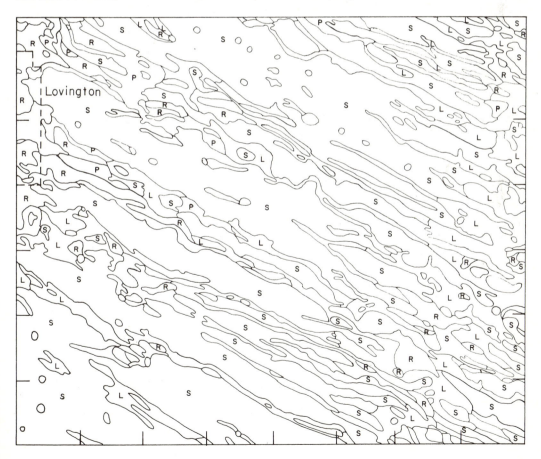

ing the longitudinal dunes at right angles, the calcrete surface is corrugated.

In Alberta, Canada, U-shaped dunes indicate effective wind directions (Odynsky, 1958). Dune alignment shows southwesterly winds in the Grande Prairie area, northwesterly winds in the Edmonton area, and southwesterly winds in southeastern Alberta (fig. 8.6). Dune orientation relates to storm tracks and surface winds. During the summer, low-pressure areas cross northern Alberta toward the southeast with accompanying northwesterly winds. During both the summer and the winter, low-pressure areas cross southern Alberta toward the northeast with southwesterly winds.

Stabilized dunes may indicate a former drier climate if other causes can be negated (H. T. U. Smith, 1949). Prior to stabilization, the conditions required for dune building were a reasonable supply of sand, winds sufficiently strong and persistent to move the sand, and lack of sufficient vegetation to hold the sand. Sand dunes may form

Figure 8.6 Orientation of U-shaped dunes in Alberta, Canada, in relation to trends of movement of atmospheric low-pressure areas and direction of surface winds. From Odynsky (1958), by permission of Agric. Inst. of Canada.

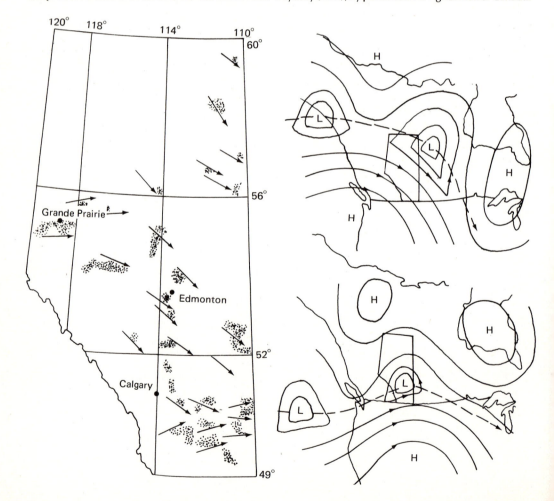

in any area in any climate if the supply of sand and wind conditions dominate vegetation. Along the Oregon coast under a mild maritime climate, active dunes encroach on coniferous forest (Cooper, 1958). Trees with shrub understory are used to stabilize the dunes by forming windbreaks and ground cover and creating a more subdued topographic surface (Brown and Hafenrichter, 1962).

Sand dunes also form where soils dry out because of a lowering of water tables by the incision of streams. Loss of ground cover by forest or prairie fire may provide dune-forming conditions. Where stabilized dunes occur in a large regional pattern, the requirement for reduced ground cover relates more readily to drier climate.

Loess

Loess is generally considered a wind-deposited sediment that is unconsolidated and is composed mainly of silt-size particles. There is an argument over the eolian origin of loess—whether it is a product of weathering (Berg, 1964) or of fluvial and colluvial processes (Russell, 1940; Fisk, 1951). But a focus on a model and a systems analysis of loess in the upper Mississippi Valley region aids in understanding the sediment and its origin.

Loess, which is usually massive in structure, is usually described as unstratified, but coarse-grained loess may be stratified and cross-bedded [fig. 8.7(a)]. Inherent in the structure of loess is its ability to stand in nearly vertical faces, and vertical cleavage is prominent [fig. 8.7(b), (d)]. The maximum stable heights of steep loess slopes relate to soil density and shear strength (Lohnes and Handy, 1968). Vertical cleavage is caused by tension in the surface layer of the loess, and failure along cleavage results in an average slope angle of about 77°.

A thick accumulation of loess causes rugged bluff topography [fig. 8.7(c)]. Such landscapes are common bordering the Missouri River Valley in Iowa, Nebraska, Kansas, and Missouri and the Illinois and Mississippi River Valleys in Illinois. These valleys are major source areas for loess, and distribution patterns relate to them.

Spatial relations

In the midcontinental United States loess forms a surface blanket of tens of thousands of square miles from the Rocky Mountain front eastward to the Appalachian Mountains (fig. 8.2). It extends down areas bordering the Mississippi River Valley almost to the Gulf of Mexico. It occurs widely in the Palouse region in Washington and adjacent Oregon and in the Snake River country of southern Idaho. Loess is a common sediment in other parts of the world as well, including northern Europe, Russia, China, and Argentina. A widespread clayey loesslike deposit in Australia is *parna* (Butler, 1956). Loess is also common in New Zealand (Cowie, 1964).

Prime evidence of the eolian origin of loess is its deposition on pre-existing land surfaces with a wide range of altitude. In southwestern Iowa, loess crosses all watershed divides, including the Mississippi-Missouri divide. Across 54 consecutive ridges and a few intervening terraces in Pottawattamie and Cass Counties, the base of the loess has an elevation range of 320 feet. Beneath the loess at most sites are undisturbed buried soils, and above the loess at all sites is the atmosphere. Silt particles had to settle through the air to bury pre-existing soils at the various elevations on ridges and terraces (Ruhe, 1969a).

In an area of divides and ridges, loess systematically decreases in thickness as the distance from the Missouri River Valley increases. Along a straight-line traverse from Sioux City, Iowa, southeastward to the center of the Missouri state line, the thinning of loess is expressed by the equation:

$$T = 1250.5 - 528.5 \log D$$

where T is thickness in inches and D is distance in

Figure 8.7 Loess in western Iowa. (a) Cross-bedded loess between "dark-colored" bands. (b) Massive loess with vertical cleavage above buried soil A horizon in loess marked by hammer. (c) A constructional landscape of loess bluffs east of the Missouri River Valley. (d) A road cut in loess with typical near-vertical face and bands.

miles (Hutton, 1947). The loess thins from 680 to 95 inches in a distance of 168 miles. If a curved traverse is laid out to cross loess thickness contours at right angles but along the same general transect, thinning is expressed by a different equation:

$$T = \frac{1}{(6.27 \times 10^{-4}) + (6.38 \times 10^{-5}D)}$$

The first equation relates thickness to the logarithm of distance; the second equation is a hyperbolic relation of the reciprocal of thickness to distance, or $1/Y = a + bX$.

This model of thinning applies in other parts of the midcontinent region. In Illinois the loess thins southeastward with distance from the Mississippi River Valley and the Illinois River Valley, as expressed by a logarithmic relation, $T = a - b \log D$ (G.D. Smith, 1942), or by an additive exponential function, $T = ae^{-bD} + ce^{-dD}$ (Frazee et al., 1970). In Indiana southeastward from the Wabash River Valley, thinning with distance is expressed by the logarithmic function (Caldwell and White, 1956).

Systematic thinning also applies to older buried loesses (Ruhe, 1968). In southwestern Iowa, Wisconsin loess buries Loveland loess. At a sequence of sites away from the Missouri River Valley, thicknesses of both loesses were measured in the same vertical sections. The lower loess thins from 26 to 16 feet in a distance of 18 miles, the upper loess from 38 to 23 feet. The respective relations are:

$$Tl = \frac{1}{(4.3 \times 10^{-2}) + (1.04 \times 10^{-3}D)}$$

$$Tu = \frac{1}{(2.8 \times 10^{-2}) + (4.7 \times 10^{-4}D)}$$

Although the absolute values differ, the rates of thinning are similar, and the sedimentological systems are alike. Universality of the model is evident.

As the loess thins, particle size systematically changes with distance. In southwestern Iowa between 13 and 64 miles from the Missouri River Valley, coarse silt Sc (62-16 μ) decreases, fine silt Sf (16-2 μ) increases, clay Cl (less than 2 μ) increases, and median particle diameter Md decreases with distance D as shown in the equations (Ruhe, 1969a):

$$Sc = \frac{1}{(1.56 \times 10^{-2}) + (6.64 \times 10^{-5}D)}$$

$$Sf = \frac{1}{(4.91 \times 10^{-2}) - (3.59 \times 10^{-5}D)}$$

$$Cl = 17.39 + 0.15D$$

$$Md = \frac{1}{(4.75 \times 10^{-2}) + (2.01 \times 10^{-4}D)}$$

There are similar particle-size distributions in the systems in Illinois and Indiana.

A decrease in bed thickness and particle size with increasing transport downwind is the product of *eolian differentiation* (Scheidegger and Potter, 1968). Particles segregate by fall velocity as they are transported downwind. Size and density are factors in fall velocity, and wind direction, velocity, turbulence, and height of trajectory of particles govern the fallout pattern.

Assuming (1) constant wind velocity, (2) ejection of particles into the air during a time interval, (3) an initial distribution of sizes and densities, (4) turbulent air decaying downwind, and (5) fallout of particles with decay of turbulence, the pattern of particle size and thickness may be expressed at various distances from the source (Scheidegger and Potter, 1968). The equation for the approximate modal particle size (Wm) of a frequency distribution at a site downwind is:

$$Wm = \left(\frac{1}{Ch}\right) \left(\frac{v^m}{x^m}\right)$$

where C is a constant; h, height above ground; v, average wind velocity; x, distance from source; and m, the turbulence decay constant. The closer the site is to the source, the smaller x is, and the larger Wm is. Particles are graded laterally, since the coarsest and densest particles fall out first.

The total thickness (H) of a bed at a site downwind is expressed by the equation (Scheidegger and Potter, 1968):

$$H = Ta \ (\frac{v}{x})^{bm + 1}$$

where T is the time interval; a and b, constants; and v and x, average wind velocity and distance. The closer the site is to the source, the smaller x is, and the thicker H is. If the time interval is longer, the bed is thicker.

The areal, altitudinal, and internal properties of loess emphasize its eolian nature, and its stratigraphic relations support this conclusion.

Stratigraphic relations

The stratigraphic study of loess is generally approached from two points of view. In one, loess is considered as one body; in the other, the increments of a loess are studied and analyzed. The first approach assumes loess to be one thinning wedge of sediment, with uniform rates of deposition at any specific location. This assumption has led to certain difficulties in explaining loess-derived soils (Fehrenbacher, 1973; Kleiss, 1973; Ruhe, 1973).

Where loess is examined in detail, units are stratigraphically separable and portray a different picture of the loess-sedimentation model [fig. 8.7(b), (d)]. In southwestern Iowa, loess is separated vertically by "dark-colored bands" (Daniels et al., 1960), and the organic carbon of the bands and a basal soil have been dated. At 7 miles from Missouri River Valley, the basal soil is 24,400 years old. Bands at 12 feet higher, another 35 feet higher, and an additional 7 feet higher are 22,350 years, 17,130 years, and 15,300 years old respectively. Ground surface 9 feet higher is about 14,000 years old (Ruhe et al., 1971). Between these stratigraphic levels, the rates of

deposition are 1 foot in 171, 149, 261, and 144 years respectively.

At 25 miles from the source, the average age of a 12-inch thick A horizon soil at the base of the loess is 22,475 years old (Ruhe, 1969b). At 7 miles from the source 12 feet of loess were therefore deposited in about 2000 years (24,450 − 22,350 = 2100 years), but at 25 miles from the source only 1 foot of loess was deposited during the same time. A lower loess increment wedges out at about 25 miles. At 7 miles, 51 feet of loess are less than 22,350 years old, but at 25 miles, 27 feet of loess are less than 22,475 years old. The lower loess wedge is overlapped by younger loess away from the source, and still younger increments, separated by other dark bands in the vertical section near the source, must be the main loess deposit farther away.

The age of the base of the loess decreases with the distance from the source, as expressed by:

$$A = 24,750 − 45D$$

where A is age in radiocarbon years and D is distance in miles. This equation applies to the curved traverse whose thinning is expressed by the hyperbolic function.

Overlapping of loess increments also occurs in Illinois, where Peoria loess contains four vertical zones with different clay-mineral assemblages (Frye, Glass, and Willman, 1968). Basal Zone I has high montmorillonite and low illite content. Zone II has less expandable clay minerals and more illite. Zone III has even less expandable clay minerals and additional illite. The uppermost zone, IV, has high montmorillonite and low illite content.

These mineralogic zones have been traced away from major valley sources (Kleiss, 1973). At a distance of 52 miles, where Peoria loess is 75 inches thick, 80 percent is composed of the youngest increment, Zone IV.

These relations are validated by other stratigraphic checks. In east-central Illinois about 10 inches of loess are under glacial drift and about

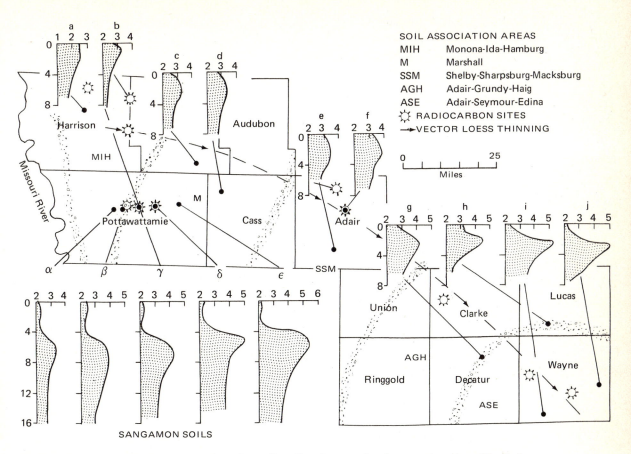

Figure 8.8 Increased soil development, as shown by profiles of less than 2 μ clay along a vector of loess thinning in southwestern Iowa. Vertical profile scale: times 10 in. Percent clay: times 10. Note the developmental sequence in Sangamon paleosols in the telescoped distance. From Ruhe (1969b), by permission of University of Alberta Printing Dept., Edmonton.

40 inches are on the drift (Fehrenbacher et al., 1965). Since wood from the till is 18,500 to 19,200 years old, 80 percent of the loess must be younger. Near the source, Zone I loess is as old as 22,000 years (Frye, Glass, and Willman, 1968).

Loess-derived soils

The loess province of the upper Mississippi River basin provides a unique model for soil-geomorphology studies. As the loess thins and particle size decreases from a source, soils formed on

the loess have progressively greater development. In southwestern Iowa clay content in the *B* horizon of the soils systematically increases (fig. 8.8), and soil patterns are in bands paralleling the valley source. Similar soil patterns and relations are in loess systems in Illinois (Fehrenbacher, 1973), Kansas (Hanna and Bidwell, 1955), and Missouri (Shrader et al., 1953).

Stratigraphic and topographic factors also change in the model. With thinning loess, particle-size decreases, and greater soil development, depth to more impermeable substrata decreases, relief decreases, and ridges usually broaden (Ruhe, 1969a). Within this framework three explanations have been given for increased soil development. (1) Parent-material effect: finer-textured thin loess permits more intensive soil formation than coarser-textured thick loess regardless of time (Hanna and Bidwell, 1955). (2) Wetness effect: in thin loess shallower depth to more impermeable substrata on broad ridges provides a wetter weathering environment than greater depth in thick loess regardless of time (Bray, 1937; Ruhe, 1969a). (3) "Effective age" idea: during the same loess-deposition time but with uniform rate at any site, deposition was slower in thin loess than thick loess. Weathering progressed during and after deposition in thin loess and provided a head start over weathering in thick loess which only followed deposition (G. D. Smith, 1942; Hutton, 1947). Combinations of the three explanations have also been used.

The basic assumptions of the first explanation are that the amounts of clay in the loess or soil C horizons are sedimentological but that the amounts of clay in the soil B horizons are sedimentological and pedogenic (Ruhe, 1969b). To analyze the system, plot the amount of clay (Cy) of each horizon against distance D (fig. 8.9). For Kansas soils the relations for both soil B and C horizons are expressed as $Cy = a + b \log D$. For B and C horizons, curves are nearly subparallel, but constants a and b differ, a being 22.237 and 14.289 and b being 21.354 and 19.149 respectively. The difference between the a constants for the B and C horizons ($Ba - Ca$) is the pedogenic impact exclusive of sedimentologic cause. The difference between the b constants for the B and C horizons ($Bb - Cb$) means that rate of change in B horizons is greater than rate of change in C horizons. A differing rate of change could be another pedogenic effect additive to difference expressed by a constants. At any site, pedogenic effect is expressed, then, as $(Ba - Ca) + (Bb - Cb)$.

This argument favors parent-material impact, but is there cause and effect? Let us examine from an alternative view the specific changes from site to site in the developmental sequence. If clay content in the B horizon at site 2 is $C2$ and at site 1 is $C1$, then the specific change is $C2 - C1$, and so on. The changes in the C horizons can be calculated in the same way. If B-horizon changes are plotted against paired C-horizon changes, no systematic relation is seen, but there would be if changes in parent material caused changes in soil development. Let us turn to the other explanations.

The idea of "effective age" is no longer tenable. The assumption that loess deposition is uniform at any site is opposed by the fact that there are variable rates of deposition and the fact that the youngest loess increment is soil parent material in the thin-loess areas (Ruhe, 1969a; Kleiss, 1973).

Wetness seems to be the most logical explanation, and it is shown by the kinds of soils that form in thick and thin loess. Well-drained Prairie or Gray-Brown Podzolic (forest) soils are formed where loess is thick, more poorly drained Humic Gley soils and Planosols where loess is thin.

Volcanic ash

Another widespread eolian sediment is *volcanic ash*, which is uncemented pyroclastic material consisting of particles less than 4mm in diameter. Like loess, volcanic ash decreases in bed thickness and particle size downwind from its source. The source of volcanic ash is a volcanic vent.

About 6600 years ago ejecta from Mt. Mazama (Crater Lake), Oregon, formed an elliptical fallout pattern northeastward through Oregon, Washington, Idaho, western Montana, southern British Columbia, and as far as Edmonton, Alberta (Westgate et al., 1969). The ash is 10 feet thick downwind at 29 miles, 7 feet thick at 34 miles, and about 2 feet thick from 67 to 278 miles from Crater Lake (Harward and Youngberg, 1969).

If a single bed of ash can be distinguished and mapped regionally, it is a key bed in stratigraphy.

One bed represents one volcanic ejection into the atmosphere at one time, followed by downwind deposition. Throughout the fallout region, any pre-existing material was buried by ash during that specific time. Any material that then buried the ash dates from that time. These are basic principles of *tephrochronology*. *Tephra* are volcanic ejecta; tephrochronology is a chronology based on the dating of volcanic-ash layers.

The problem, then, is to determine whether there is one or more than one ash. A given ash usually has specific chemical and petrographic characteristics (Wilcox, 1965), and the latter include shape, internal structure, degree of alteration, and refractive indices of shards (Swineford and Frye, 1946). *Shards* are curved, spiculelike fragments of volcanic glass. Instrumental neutron-activation analysis is also useful for distinguishing different ash deposits (Harward and Youngberg, 1969). Among 21 nuclides that characterize tephra in Oregon, many of them distinguish Mazama from Newberry, St. Helens, and Glacier Peak ashes.

When a specific ash is distinguished, the bed is traced cross-country to a source vent. Early Pleistocene Pearlette ash was traced from the glaciated areas of western Iowa to the Valle Grande volcanic vent in north-central New Mexico (Swineford, 1949). But at present there is doubt as to the validity of the source (Wilcox, 1965) and whether the Pearlette is only one ash.

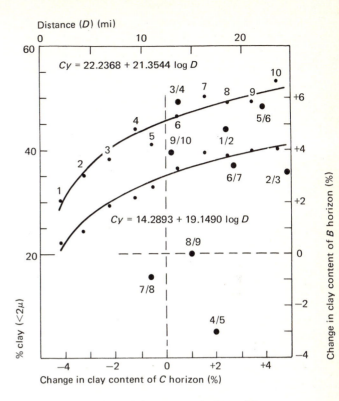

Figure 8.9 The relation of clay content of soil B and C horizons in northeastern Kansas to the distance from the source and the relation of change in clay content of soil horizons between successive sites. Calculated from Hanna and Bidwell (1955). From Ruhe (1969b), by permission of University of Alberta Printing Dept., Edmonton.

Aerosols

Aerosols are particles dispersed in a gas, and those carried in the atmosphere may cause rather unusual phenomena upon deposition in rainfall. Recall the sodium and chloride contents in soils in San Mateo County, California (Chapter 3). Among the ions carried in atmospheric moisture over the United States, the greatest concentration of Ca^{++} is over the Great Plains and southwestern deserts, where dust storms are most prevalent (Junge and Werby, 1958). In southern New Mexico many soils formed in parent materials with low calcium content but have thick carbonate horizons. There is not enough Ca^{++} in the parent material that can be released during weathering to form the $CaCO_3$ of the K or ca horizons (Ruhe, 1967). Carbonate can be introduced by calcareous dust falling on the soil surface, then moving downward in solution in a wetting front and precipitating (table 8.2). Carbonate may also be introduced by Ca^{++} in rainfall penetrating the soil and combining to form the carbonate (Gile et al., 1970).

Soils of the Hawaiian Islands contain quartz although unaltered rocks there do not (Rex et al., 1969). Almost 70 percent of the quartz particles are in the 10 to $2\,\mu$ size fraction that is characteristic of aerosols and tropospheric dust. The concentration of quartz in surface horizons increases functionally with land elevation and correlative rainfall. The extraneous source of quartz must be the atmosphere and, for the Hawaiian Islands, trans-Pacific circulation. This is an extreme example of the lack of geographic limitation on eolian processes affecting landscapes and soils.

Up to this point attention has been focused on running water in channels and on hillslopes, the processes involved, and the effects on land. Now we will consider water in a basin and its effects on the land around it in terms of erosion, transportation, and deposition caused by waves and currents. Since ice also affects shorelines, solid-phase phenomena will be examined as well.

Waves are generated by wind, earth disturbances, movement of an object like a boat across water, and the gravitational pull of the moon and the sun. Turbulent flow of air over the water surface creates a tangential stress forming ripples and waves. Earthquake sea waves, or *tsunami*, are caused by the rapid motion of underwater rocks disturbing the overlying water mass. The water surface oscillates, and a series of seismic sea waves move outward from the site of the disturbance. Tidal waves are long waves caused by revolution of the earth within the gravitational fields of the moon and the sun. Bulges of water are pulled from the earth by gravitational force, which causes low tide at the shoreline. With rotation of the earth and reduced gravitational force, the bulges flatten, which causes high tide.

Wave action

When waves are generated by wind, kinetic energy is transferred from wind to wave to land. The dynamics of this system have been thoroughly studied and treated in technical reports of the Coastal Engineering Research Center (CERC),

⑨

Shore Processes and Features

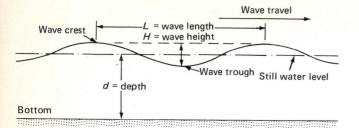

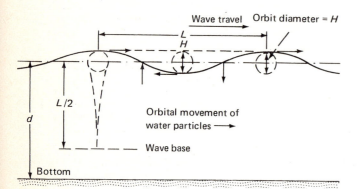

Figure 9.1 Wave characteristics and water-particle movement. From Jachowski (1966).

U.S. Army, Corps of Engineers (Jachowski, 1966; Allen, 1972). Its material will be used liberally in the following discussion.

Wave dynamics

In deep water, waves formed by wind are oscillatory, and water particles move in a circular pattern about a mean position (fig. 9.1). An idealized wave is a sine curve, expressed by $Y = \sin X$. The wave length (L) is the horizontal distance between two successive wave crests. The wave height (H) is the vertical distance between a wave crest and its adjacent trough. The wave period (T) is the time required for two successive wave crests to pass a given point. The equation for wave velocity (C) is:

$$C = \frac{L}{T}$$

Wave velocity is related to wave length and water depth in the equation:

$$C^2 = \frac{gL}{2\pi} \tan h \frac{2\pi d}{L}$$

where g is the acceleration of gravity and d is the water depth measured from the stillwater level (X the axis of the sine curve) to the bottom. When d is large compared to L, the hyperbolic function ($\tan h \frac{2\pi d}{L}$) is almost one, and wave velocity is independent of depth in deep water, or:

$$C^2 = \frac{gL}{2\pi}$$

If $L = CT$

$$C = \frac{gT}{2\pi} = 5.12T$$

$$L = 5.12T^2$$

where deep water depth is defined as depth greater than half the wave length.

When d is small compared to L, the hyperbolic function almost equals $2\pi d/L$ and wave velocity is expressed:

$$C^2 = gd$$

$$C = \sqrt{gd}$$

Shallow water depth is 1/25 the wave length or less, and transitional water depth is between 1/2 and 1/25 the wave length.

Wave height is a function of wind velocity, the *fetch* (the distance across the water that the wind blows), and the length of time that the wind blows. On small lakes a more direct relation exists between wind velocity and wave height than between wave height and fetch (fig. 9.2).

Wave height is a factor in wave energy, and total energy (Et) in a single wave of unit width is determined by the equation:

$$Et = \frac{pgH^2L}{8} (1 - M \frac{H^2}{L^2})$$

where p is the mass density of water and M is an energy coefficient. For deep-water waves Et is approximated:

$$Et = \frac{wLH^2}{8}$$

$$Et = 8LH^2$$

where w is water weight of 64 pounds per cubic foot. Et is measured in foot-pounds per unit width of wave.

Total energy consists of kinetic and potential energy. Kinetic energy depends on the water-particle velocity within the wave motion. Potential energy is represented by the water mass lifted from the wave trough and standing above the still-water level beneath the wave crest. Water particles follow a circular orbit in deep water and an elliptical path in shallow water (fig. 9.1). Below a depth of half the wave length, generally known as the *wave base*, orbital motion is minimal.

In deep water about half the total energy in each wave is expended in moving the wave forward and sustaining the wave train. The remaining energy can be expended on shore and is also about half the total energy of the deep-water wave. The total energy expended on the shore during a given time is the product of the single-wave energy and the number of waves reaching shore during that time.

As waves approach the shore the water depth and the wave velocity decrease. If the wave period remains constant, the wave length also decreases with depth. These *shoaling* changes occur in transitional and shallow water. The wave height increases until a wave becomes unstable, peaks, and breaks. The orbital water motion changes to turbulent surf conditions, and water is discharged toward the shoreline (fig. 9.3).

Wave patterns

The variation of wave velocity with water depth affects the wave pattern. The part of the wave in deeper water moves faster than the part in shallow water, and the wave crest bends to align with underwater contours, which causes *wave refraction* (fig. 9.4). Not much energy is transmitted

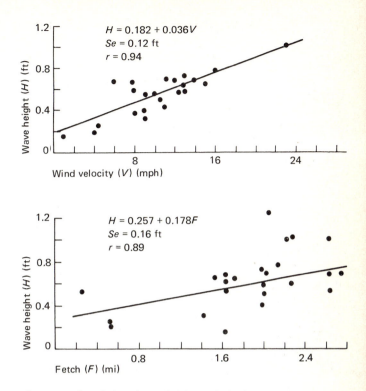

Figure 9.2 **The relation of wave height to wind velocity and fetch on the Iowa Great Lakes.** From Miller (1971), by permission.

laterally along the crest, but energy remains constant between *wave rays*, or *orthogonals*, which are perpendicular to the wave crests as they pass over changing underwater slopes.

As the wave crests align with underwater contours during shoaling, the orthogonals change direction, as approximated in the formula:

$$\sin a_2 = \frac{C_2}{C_1} \sin a_1$$

where a_1 is the angle between a perpendicular to an orthogonal and an underwater contour; a_2, a similar angle measured at the next shoreward contour; and C_1 and C_2, respective wave velocities.

Other patterns are *wave diffraction* and *wave reflection*. Diffraction is caused by a train of

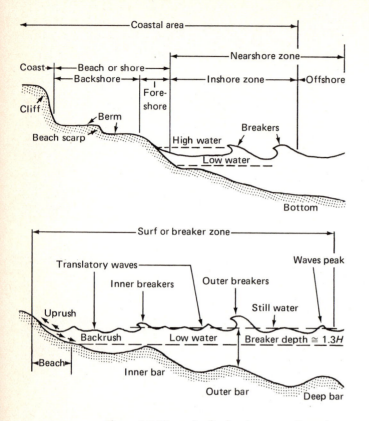

Figure 9.3 Waves in the breaker zone. From Jachowski (1966).

waves bending around the edge of an obstacle. During diffraction, energy is transmitted laterally along a wave crest. Wave reflection is caused by the approach of a wave at an angle to a shoreline and is analogous to the reflection of light or sound waves. The incident and reflected wave rays form equal angles with perpendiculars to the shoreline (fig. 9.8).

Water-level changes

While waves are acting and patterns are changing, water levels are also changing. *Tide* is the periodic rise and fall of water level caused by the gravitational attraction of the moon and the sun

on the rotating earth. The highest tides, *spring tides,* occur when the sun, moon, and earth are aligned with the moon either toward or away from the sun relative to the earth. The lowest tides, *neap tides*, result when the sun and moon are at right angles with respect to the earth. A *semidiurnal* tide has two high and two low waters during a tidal day; a *diurnal tide* has only one high and one low water during the same period (fig. 9.5). A combination of them is a *mixed tide*. The declination of the moon from a position directly over the earth's equator causes diurnal and mixed tides. The water levels associated with tides are termed lower or higher low water and lower or higher high water (fig. 9.5). The mean ranges of tidal levels may vary with latitude. Along the East Coast, they are 10 to 20 feet in Maine, 3 to 5 feet in New Jersey, and 1 foot at Key West, Florida.

Storm surges cause water-level changes as winds accompanied by atmospheric-pressure changes produce a leeward piling up and a windward lowering of water. Examine the effect of Hurricane Donna across the Florida Keys and Florida Bay in September, 1960 (Conover, 1960; Gleason, 1972). The storm crossed the axes of the keys and bay at right angles and passed over Cape Sable (fig. 9.6). Water level was reduced to 5 feet below mean sea level in the eastern part of the bay but welled up 11 feet above mean sea level near Flamingo, which is only a few feet above sea level. That land area was inundated. To help avoid catastrophes in populated areas, storm surges and surge amplitudes are predicted for given meteorological conditions (Welander, 1961).

A *seiche* (pronounced sāsh) is an oscillatory wave of a relatively long period that continues after cessation of the originating seismic or atmospheric force. Seiches occur in bodies of water such as lakes, canals, and bays. They are caused by sudden changes in atmospheric pressure, wind velocity, earthquakes, underwater landslide, or sudden addition or subtraction of water in a canal or through the mouth of a bay. In 1968 a tornado crossed Lake Okoboji, Iowa, and eye witnesses observed a doming of the lake level within the low-pressure eye of the tornado. Water withdrew rapidly from the shoreline and suddenly returned in oscillatory waves above mean water level as the storm passed onto adjacent land.

Stage changes are ordinary fluctuations in the hydrologic budget of a water basin. Variation in rainfall, whether seasonal or longer-cycle, causes a change in water level (fig. 9.7). Evaporation caused by higher temperatures or lower humidity or both result in lower water levels, and these changes may be seasonal or longer-term.

Basin changes cause differences in water level. If leakage to the subsurface dominates recharge, water level lowers. In karst terrain, a sink hole may unplug and the water drain completely, much like a bathtub after the plug is pulled. Headward extension of a stream may breech the basin perimeter, lower the water level, and drain the lake.

All of these changes in water level show that shore processes may be active at various topographic levels along a shoreline. They must be considered carefully when the historic implications of a sequence of shorelines are being evaluated.

Currents

Currents form near shore as a result of discharge from successive breaking and translatory waves. The beach is an obstacle that reduces discharge onshore to zero. Excess water piles up, which causes superelevation of the water surface or an *elevation head,* and water must flow away from the locus of the head.

If waves approach normal to a shoreline, water returns in the opposite direction, forming a current. The flowing water follows underwater surface irregularities and may become channeled, forming a *rip current.* The channel may erode and be supplied by feeder lines parallel to the beach.

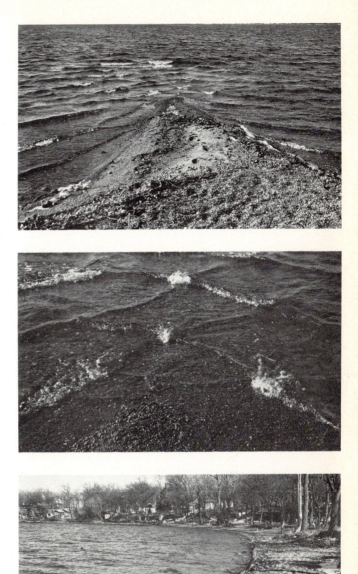

Figure 9.4 Wave refraction at a point and along the shore. Photographs by G. A. Miller, by permission.

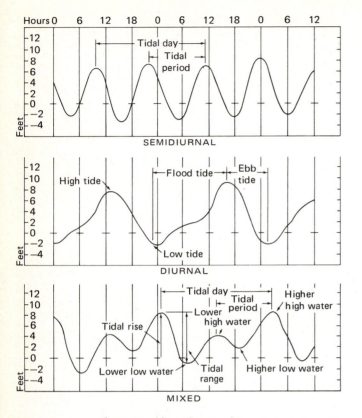

Figure 9.5 Tides and nomenclature. From Jachowski (1966).

The rip current flows under and beyond the breaker zone and effectively increases the wave velocity above the current. Laterally along the same wave velocity is not increased if rip currents do not exist. Wave and current action accordingly becomes more complicated.

If waves approach a shoreline at an acute angle, a *longshore current* forms which parallels the shore (fig. 9.8). The wave crest (*W*) approaches the shoreline (*S*) at an incident angle (*i*). Points along the wave crest strike the shoreline at successive times of arrival: *a, b, c, . . . h*. The shoreline reflects the wave at angle *r*, and the position of the reflected crest *R* at time of impact at *h* is approximated by a tangent to arcs whose radii are aa', bb', . . . gg'.

At each station, water moves onshore at angle *i* with resultant backrush at angle *r*. But backrush at *b* is intercepted by uprush toward *c* and so on, and a zigzag pattern of uprush and backrush develops. The longshore current is the resultant *L . . . L'* of the vectors *i* and *r*.

Waves and currents cause discharge of water near shore and onshore, and, like running water (Chapter 3), are responsible for littoral processes of erosion, transport, and deposition of sediment.

Littoral processes and shore features

Waves first contact the bottom where water depth is about equal to half the wave length, and consequently little erosion or sediment transport takes place at greater water depth. In reality, no appreciable erosion or transport occurs until the water depth is much shallower. Erosion and deposition create features along the shore. In the following sections, the definitions of features will follow those of the Coastal Engineering Research Center (Allen, 1972).

Erosional forms

The processes whereby rocks and sediments are eroded along a shore are the well-known corrosion, corrasion, and hydraulic-action processes (Chapter 3). Water can dissolve soluble materials and may be very effective on calcareous rocks and sediments. Corrasion involves the direct effect of waves containing mineral particles. Rocks may be worn away, smoothed, and polished by the abrasive action of the water-mineral mix. Hydraulic action is the impact of water, which can cause rock breakdown. When a wave impacts a rock face, the air in the cracks is compressed. When the wave recedes, the air suddenly expands. These sudden changes in pressure can enlarge the cracks and cause material to be pried from the face. Pressures may range from half a ton to many tons per square foot.

These processes erode rock and sediment and

ultimately form a steeply inclined face. If this feature has low relief, it is a scarp; if it has high great relief it may be a *sea cliff,* which is a cliff at the seaward edge of a coast.

Wave action causes undercutting of a steeply inclined shore, creating *niches,* or *nips,* which are indentations in the rock face at or near water level. Overhanging rock masses may break off, which causes the shoreline to recede. *Caves* and *arches* are commonly associated with niches.

As the cliff or scarp retreats, a wave-cut bench or *platform* is eroded in front of the cliff or scarp. The retreat of the cliff is genetically linked with the cutting of the platform. This retreat requires wave action directly on shore rock and lack of a protective beach (which would prevent such action). The erosion remnants standing above water level on the platform and left behind the retreating cliff are called *stacks.*

Depositional forms

When waves and currents act on a shore, they may aggrade it as additions of sediment dominate erosion. The shore is stable when erosion and accretion are balanced; it erodes when removal of sediment dominates accretion.

Sediment is supplied by the wave erosion of adjacent shore areas and shore rocks and by streams emptying into the basin. Sediment is moved as *beach drift* and is carried in a zigzag pattern by the uprush and backrush of waves approaching at an angle to the shoreline. The suspended load is moved in longshore currents and breaking waves in the surf zone. The bed load is moved along the bottom by sliding, rolling, and saltation within and seaward of the surf zone.

The energy of waves and currents required to move particles depends on velocity, and dynamics similar to those of running water are involved (fig. 3.11). The velocity determines the particle size that is eroded, transported, and deposited. Velocity differentials determine size sorting, which can be along the beach or along the beach slope.

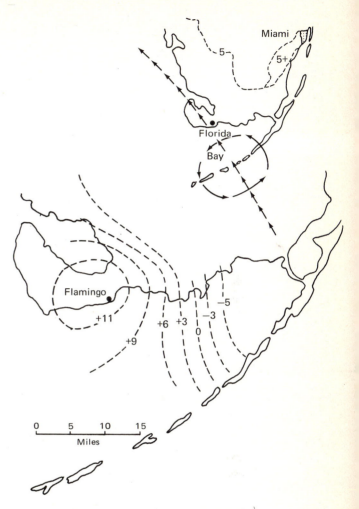

Figure 9.6 The path of Hurricane Donna across the Florida Bay in 1960. From Perkins and Enos (1968), by permission of University of Chicago Press. The magnitude of the storm surge at Flamingo. From Conover (1960), reported in Gleason (1972), by permission.

Sediment is lost from the littoral system by transport of material laterally from an area, offshore into deep water, and into underwater topographic depressions, and by wind transport of beach sediment inland. During storm surges, sediment can be transported and deposited in backshore areas and lagoons. During Hurricane

Donna in southern Florida, about 5 inches of calcareous mud were deposited as far inland as 2 miles from the Florida Bay (Gleason, 1972).

The major features formed by deposition along a shore are bars, beaches, and spits. A *bar* is a submerged or emerged embankment of unconsolidated material built on the bottom in shallow water by waves and currents. One kind of bar is a *barrier beach,* which is essentially parallel to the shore and whose crest is above the normal high-water level. Barrier beaches also are known as *offshore barriers* or *barrier islands.* A *baymouth bar* is one that extends partly or entirely across the mouth of a bay. A *cuspate bar* has a crescent shape with each end connected to the shore. A *tombolo* is a bar that connects an island to the mainland or to another island.

Bars commonly occur along sandy beaches and form on the bottom near the breaker point (fig. 9.3; Shepard, 1950). Usually there is more than one bar, and troughs intervene between

them. Near the breaker point, net sand transport is shoreward. The bars apparently are built by sediment thrown shoreward by the breaking waves. Sand builds up in ridges slightly seaward of the plunge point, and the plunging breakers excavate the bottom and produce the troughs. The movement of sand for bar construction relates to the wave height and the wave period (C. A. M. King, 1972). Near the breaker point, the amount of sand transported increases with increasing wave height for a given wave period. As the period increases for a given wave height, the volume of sand moved also increases.

A *beach* is the zone of unconsolidated sediment that extends landward from the low-water line to the place where there is a marked change in material or physiographic form, or to the line of permanent vegetation. The seaward limit is the mean low-water line. A beach includes the foreshore and the backshore (fig. 9.3). The *foreshore,* or *beach face,* is the part of the beach

Figure 9.7 Mean annual stages of Spirit Lake, Iowa, with annual rainfall for almost a century of records. From Miller (1971), by permission.

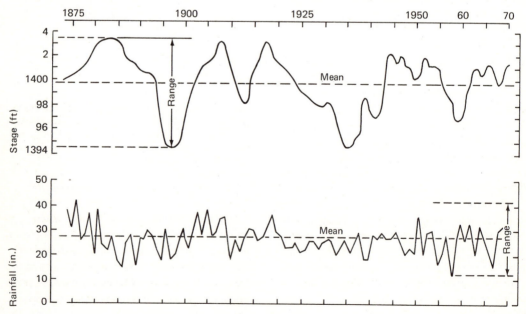

that is between the crest of the seaward berm and the ordinary low-water mark and that is ordinarily crossed by uprush and backrush as tides rise and fall. The *backshore,* or *backbeach,* is the part of the beach that is between the foreshore and the coastline and that is acted upon by waves only during severe storms with high water. It includes one or more berms. The *beach berm* is the nearly horizontal part of the beach or backshore that is formed by deposition of material by wave action. The *berm crest* is the seaward limit of a berm.

Another feature of a beach is a *beach ridge,* which is an elongate mound of beach sediment shaped by wave or other action. Beach ridges may occur singly or in parallel sets. A *beach cusp* is a crescent-shaped trough spaced at more or less regular intervals along the beach face. A *beach scarp* is a steep, nearly vertical slope along the beach that is caused by wave erosion. A *shingle beach* is a beach composed of smooth, flat or rounded pebbles and cobbles that are arranged like shingles on a roof.

There are changes in the processes affecting beaches. The processes may degrade during the winter because of higher tides and larger and steeper waves, and they may aggrade during the summer (Yasso, 1971). Shorter cycles also occur. On the northwest coast of the Gulf of California, higher levels of spring tide are reached during early afternoon and coincide with the strongest onshore sea breezes. The position of the beach berm follows the elevation of the higher high water in a two-week cycle (Inman and Filloux, 1960). At Sandy Hook, New Jersey, the uprush-backrush and breaker zone shifts up and down the beach during the semidiurnal tidal cycle. With each rising tide, about 0.02 foot of sand is deposited. About 0.2 foot of scour follows, and it in turn is followed by deposition of 0.5 to 0.7 foot of sand. With each falling tide, the changes are repeated in reverse order, and the beach is restored to its original elevation, slope, and composition (Strahler, 1966). An "equilibrium" beach profile is thus established.

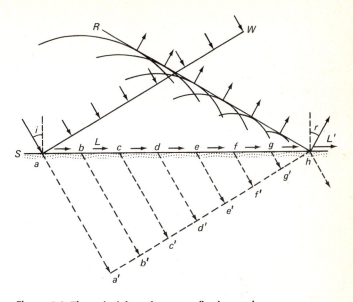

Figure 9.8 The principles of wave reflection and formation of a longshore current.

Longshore currents are particularly important in the construction of a shore feature such as a *spit,* which is a small point of land or a narrow shoal projecting into a body of water from the shore (fig. 9.4). Spits may have linear or curved lengths, and curvature may be described by circular arcs or logarithmic spirals (Yasso, 1971). If the end of spit curves landward, the form is termed a *hook.*

In the construction of these features, the amount of longshore transport depends on the longshore component of wave power, which in turn depends on the angle of wave approach. These features commonly form where there is a re-entrant along a shore that results in an abrupt change in shoreline direction. Wave refraction around this point causes a reduction in wave energy, and deposition occurs at the point and extends into the water of the re-entrant, forming a spit. The angle of incidence of the waves increases as the shoreline changes direction into the re-entrant, so the rate of longshore movement should increase, causing the spit to curve inward,

Figure 9.9 (a) A contemporary wave-cut platform on basalt with tidal pools. (b) Contact of the Ewa Coastal Plain with the Waianae Mountains, Oahu, Hawaii.

and causing a hook to form (Zenkovitch, 1967). Spits and hooks may form along any shoreline where obstructions interrupt the longshore flow.

Biogenic forms

Reefs are offshore consolidated-rock hazards to navigation at depths of 10 fathoms (60 feet) or less, and many kinds of reefs are formed by organisms. Corals, which are major reef builders, are marine coelenterates that occur as solitary or colonial polyps. The corals that form large reefs are limited to warm, shallow waters, and they occur in association with other organisms such as algae and bryozoans. These organisms build reefs by forming external skeletons dominantly of aragonite and calcite. These reefs are generally known as *coral reefs*, even though other organisms are involved. (The term *coral* also applies to the mineral material of the reef.) If a reef is attached to the shore, it is a *fringing reef*. If it is parallel to but at some distance from the shore, it is a *barrier reef*. Ring-shaped coral reefs, which have low sand islands and enclose a body of water, are *atoll reefs*, and the entire form is known as an *atoll*.

The only living coral reefs off the mainland United States are along the outer arc of the Florida Keys (Moore, 1972), where coral and associated organisms live in warm, shallow water to depths of about 26 feet. The corals have common names such as head, brain, elk horn, stag horn, and lettuce, which are easier to remember than *Monastrea annularis* (head coral) and *Diploria strigosa* (brain coral) and the like. Since corals are usually good indicators of a warm, shallow marine environment, fossil corals on land are useful geomorphic indices.

Historic extension

Three mechanisms suspend shorelines on higher land along seacoasts. Sea level may be lowered by *eustatic change*, which is a change of water level alone without vertical movement of the land mass. Eustatic change is generally associated with glacial episodes; large masses of glacier ice accumulated on the continents, causing corresponding lowering of sea level. With deglaciation, glaciers melted and water returned to the sea, raising the water level. The change could have been as much as 350 feet at the time of the last glacial maximum (Donn et al., 1962).

Sea level is also raised or lowered by *tectonic change* of the land mass relative to the sea. The land mass may be downthrown or upthrown along a fault, or the margin of the mass may be warped. *Tectono-eustatic change* is the combination of the two. Where relict shorelines are to be related to glacial eustatic sea-level changes, all possible tectonic movement of land must be ruled out or corrected for in analysis.

Changes in lake levels are caused by tectonic movement, leakage in the subsurface, drainage through an outlet, or climatic changes. Where relict shorelines are to be related to climatic

a

b

Figure 9.10 Air photographs of the Ewa Coastal Plain, Oahu, Hawaii. The distance from Barbers Point light (a) to the entrance to Pearl Harbor (b) is about 9.5 miles. The width of the plain at Pearl Harbor is 5.5 miles (b).

Figure 9.11 Relict shore features, Oahu, Hawaii. (a) A wave-cut bench on basalt 20 to 30 feet above sea level. Note the basalt-boulder shingle beach. (b) Coral rock on a Waianae Mountain slope 95 feet above sea level. (a) and (b) at Kaena Point. (c) Deltaic deposits with bottom-set, fore-set, and top-set beds topping about 25 feet above sea level. (d) An oyster bed with base about 17 feet above sea level. (c) and (d) on Waipio Peninsula.

change, all other possible causes must be ruled out or corrected for in analysis.

Relict shorelines

Relict shorelines are common features on slopes above current sea and lake levels and dry-lake floors. In the Palos Verdes Hills along the Pacific Coast southwest of Los Angeles, there is an excellent staircase of marine terraces, with 13 terrace steps from 100 feet to 1300 feet above sea level (Woodring et al., 1946). Immediately offshore there are five submerged terraces whose outer edges are 30 to 290 feet below sea level (Emery, 1958). There are five marine terraces in Santa Cruz, California; each platform was cut during rising sea level, and its cover of marine sediments was deposited during falling sea level (Bradley, 1957). Each terrace represents one oscillation of sea level.

An excellent display of relict lake shorelines is found around the Utah, Sevier, and Great Salt Lakes in Utah, Idaho, and Nevada (Morrison, 1965). The areas of the present lakes are 1000 to 2500 square miles, and the water-level elevation of Great Salt Lake is about 4200 feet. During glacial or pluvial (rainy or wet) times Lake Bonneville occupied 19,940 square miles and had a maximum depth of 1100 feet. The *Bonneville shoreline* is 1000 feet above the present Great Salt Lake. Two other prominent levels are the *Provo* and *Stansbury shorelines*, 625 and 330 feet, respectively, above Great Salt Lake.

Identification and reconstruction of relict shorelines are based on principles that are exemplified by the island of Oahu, Hawaii, where numerous old shorelines have been recognized above the present sea level (Wentworth and Palmer, 1925; Stearns, 1935a, b, 1961; Wentworth and Hoffmeister, 1939; Chapman, 1946). The shorelines and their elevations above sea level are: Manana, 2 feet; Kapapa, 5 feet; Hanauma, 12 feet; Waimanalo, 25 feet; Waialae, 45 feet; Kahuku, 55 feet; Laie, 70 feet; and Kaena, 95 feet.

If these features are discernible relict shorelines, they can be used as key levels in a sequence of geomorphic surfaces and serve as events in the chronology of landscapes. They should be examined in detail (Ruhe et al., 1965a). The 2- and 5- foot levels are not relict shorelines but are contemporary high-water platforms containing tidal pools [fig. 9.10(a)].

The Kaena shoreline has coral and calcareous beach deposits, marine clays and scarps, that are 80 to 100 feet above sea level. Benches and a coastal plain rise to those elevations. The Ewa Plain abuts the southern end of the Waianae Mountains within this elevation range (fig. 9.9), and the contact is easily mapped [fig. 9.10(b)]. At a given elevation on the plain, clay and coral were drilled to specific depths: at 110 feet they were drilled to 18 feet; at 85 feet, to 61 feet; at 75 feet, to 46 feet; and at 65 feet, to 31 feet. Coral rock is on the surface at 50 feet above sea level (Ruhe et al., 1965b). At Kaena Point [fig. 9.11(a)] coral rock is attached to the steep hillslope at an elevation of 95 feet [fig. 9.11(b)]. All of these marine deposits and landscape features above sea level demonstrate a relict Kaena shoreline.

A second prominent relict shoreline near Waimanalo Beach, Oahu, is about 25 feet above sea level. It is marked by nips in sea cliffs 22 to 27 feet above sea level. On Waipio Peninsula, deltaic deposits with typical bottom-set, fore-set, and top-set beds and oyster beds between silts and clays descend seaward below an elevation of 25 feet (fig. 9.11). At Kaena Point a bench on basalt descends seaward from the foot of the sea cliff at an elevation of 30 feet, is mantled by a dense, heavy dark-gray clay, and terminates seaward at a basalt boulder beach [fig. 9.11(a)]. The 25-foot Waimanalo shoreline is proved to be relict by marine deposits and landscape features.

Since relict shorelines marked by scarps, benches, or remnant plains can be analyzed morphometrically, 275 topographic profiles were constructed down nose slopes bounding the coast of Oahu (Ruhe et al., 1965a). Among 556 scarps and 593 benches, two distinct modes occur in the frequency distribution at 24.9 ± 7.3 feet and 92.6 ± 22.6 feet. This application of morphometric analysis supports the distinction between the Waimanalo 25-foot shoreline and the Kaena 95-foot shoreline.

The other relict shorelines presumed to be on Oahu are difficult to identify and to map in the field and were not distinct in morphometric analysis (Ruhe et al., 1965a). The two prominent relict shorelines, however, are readily identified and recognized by the association of morphometric, topographic, and lithologic features. On an island like Oahu in warm Pacific waters, coral rock at an elevation above sea level defines a sea stand higher than what now exists.

Except for marine features, similar criteria identify relict shorelines around lakes. Four times in 12 years water occupied Lake Isaacs, a playa lake in the Jornada basin in southern New Mexico, and covered about a fifth of a square mile (Ruhe, 1967). On an oblique air photograph (fig. 9.12) the older shoreline parallels the current intermittent shoreline and is marked by the outer edge of the lighter-colored photo pattern. The older basin covers about half a square mile. A still older basin, distinguished on the air photograph by a

a

b

Figure 9.12 Relict shorelines and lake basins around Isaacs Lake, Jornada Basin, southern New Mexico, looking south (a) and north (b). Air photographs by J. W. Hawley, by permission.

darker tone, extends as a narrow arm to the south and flairs outward widely to the north [fig. 9.12(b)]. The oldest basin covers about 18 square miles. These relict shorelines are readily distinguished in the field by beach scarps, beach ridges, and dunes associated with the beaches. The lake sediments are clays and silty clays in contrast to coarser-textured ridges.

Implications of relict shorelines

Shore features above the present sea level along coasts are direct evidence of former positions of the sea in relation to land. But complex problems arise: Has sea level moved up and down while land remained stable? Has the land moved up and down while the sea remained stable? Or have both land and sea been going up and down?

The solution of the first problem requires deductions from volumes of land ice during glacial times (Flint, 1971). If all existing land ice melted and returned to the sea, its level would rise about 210 feet. Conversely, during a glacial maximum more ice would be on the land, and sea level would lower about 430 feet. These values must be reduced by about one-third because of the effects of loading and unloading on the earth's crust (Flint, 1971). Sea level could have been roughly 150 feet higher or 300 to 350 feet lower (Donn et al., 1962) during interglacial and glacial times. If relict shorelines and submarine platforms are within this elevation range (and if vertical tectonic movement or warping are negated), eustasy is usually given favorable consideration.

The shorelines on Oahu, Hawaii, including the Kaena 95-foot shorelines, are considered eustatic levels (Stearns, 1961), and they are not distorted around the island (Ruhe et al., 1965a). Several submarine platforms offshore descend to 300 feet below sea level and are also considered eustatic levels (Stearns, 1961). The Lualualei platform is too deep at 1200 to 1800 feet and must be a tectonic feature as it is distorted around the island (Ruhe et al., 1965a).

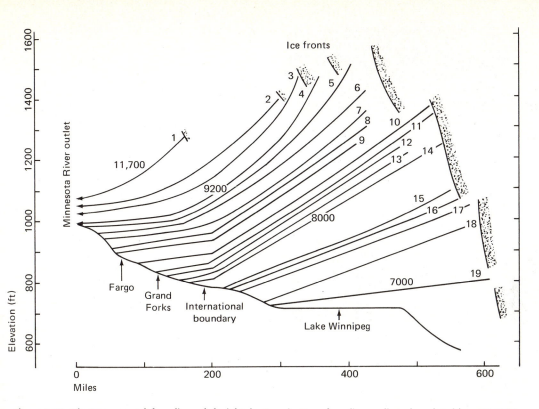

Figure 9.13 Nineteen warped shore lines of glacial Lake Agassiz. Four shore lines radiocarbon-dated from 11,700 to 7000 years ago. From Elson (1967), by permission of University of Manitoba Press, Winnipeg.

Many of the marine terraces along the southern California coast are too high to have been produced solely by eustatic change and must be related to uplift (P. B. King, 1965). Some of the lower terraces (Woodring et al., 1946) and submarine platforms (Emery, 1958) are within the range of eustatic change. A compound explanation is required for all of these marine terrace levels.

Relict shorelines around lakes in arid regions are usually explained by climatic change. After tectonic effects, subsurface leakage, and drainage through an outlet are ruled out, the problem reduces to calculating the volume of water necessary to raise lake level to a specific shoreline above an existing lake or dry bed. The volume of the current lake is calculated and is related to present rainfall, temperature, evaporation, and runoff recharge to the lake. A simple algebraic equation is set up and solved for the past climate. The major complexity is evaluating past temperature, which must be considered because evaporation from a free-water surface is dependent on it.

Consider Pluvial Lake Lahontan in Nevada, of which the present Walker and Pyramid Lakes are remnants and references (Morrison, 1965). Relict shorelines are 110, 320, and 550 feet above the present lake level. Assuming temperatures similar to those today, an annual rainfall of 20 inches would raise water level to its highest relict shoreline. Correcting for lower temperature by about 9°F, an annual rainfall of about 18 inches would raise water level to the maximum level. The current annual rainfall is about 6 inches.

a

b

c

Figure 9.14 Lake-ice phenomena at the Iowa Great Lakes during the winter of 1969-1970; pressure ridges and ice ramparts. Note the entrapped debris. The height of ice rampart in (b) is 8.5 feet. Photographs by G. A. Miller, by permission.

Relict shorelines may be deformed and not be level along a seashore or around a lake. The highest shoreline of pluvial Lake Bonneville is about 210 feet higher on relict islands near the center of the old lake basin than around the perimeter (P. B. King, 1965). The upwarp of the basin center is due to the unloading of lake water, which was about 1100 feet deep.

Warped shorelines mark old glacial lakes in the midcontinental United States. Many relict shorelines of the glacial Great Lakes are horizontal northward along the present lakes but rise in elevation farther north. A *hinge line* separates the horizontal from the rising shoreline (Wayne and Zumberge, 1965).

Warped shorelines characterize glacial Lake Agassiz, which occupied 77,000 to 193,000 square miles in Minnesota, North Dakota, Manitoba, Ontario, and Saskatchewan. Some remnants of this glacial lake are Lake-of-the-Woods, Lake Manitoba, and Lake Winnipeg. From 11,700 to 9200 years ago this lake was impounded by glacier ice to the north and drained southward through the Minnesota River Valley to the Mississippi River (fig. 9.13). During four lake stages, water discharged southward and cut the Minnesota Valley trench. With subsequent lake drainage northward, the Minnesota River was left as an *underfit stream* in the southern drainage outlet of Lake Agassiz. The lake level dropped intermittently, creating shorelines at numerous stillstands to 8000 and 7000 years ago (Elson, 1967). All of the shorelines rise in elevation northward. The warping of the shorelines of these glacial lakes was caused by crustal rebound in response to unloading by glacier ice during deglaciation.

Lake ice

An understanding of lake ice and its effects requires an appreciation of the unique properties of water. It is one of the few materials that expands when it freezes. A given volume of ice weighs less

than the same volume of water, and consequently ice floats.

Physics of the water-ice system

Water has its greatest density (1.0 g/cm³) at 4°C (above its freezing point of 0°C). The density decreases to about 0.9998 g/cm³ approaching 0°C, but upon freezing, it abruptly decreases to 0.9168 g/cm³. The volume increases by about 9 percent, and this excessive expansion causes enormous pressure on boundaries if water is frozen in a container.

Below freezing temperature, ice contracts with further reduction in temperature and expands when temperature increases. In lake ice, a reduction of temperature of only 0.15°C/hr for 16 hours can cause contraction cracks (Wagner, 1970). A temperature increase of 0.3°C/hr can cause ice expansion, and a total change of 2°C can cause 3 cm of contraction and expansion. Water freezes in contraction cracks and increases the area of the ice layer, with a net increase in expansion that causes additional stresses on the shoreline.

Two thermal properties, specific heat and latent heat of fusion, are important in the water-ice system during lake freezing, and the values of these properties are unusually high in water. *Specific heat* is the amount of heat required to

Figure 9.15 Beach ramparts formed by ice thrust at the Iowa Great Lakes during the winter of 1969-1970. Photographs by G. A. Miller, by permission.

raise the temperature of a unit substance one degree and is expressed in BTU/lb/°F or cal/g/°C. Pure water, the standard, has a value of 1.0. In contrast, the value for quartz sand is 0.2 and that for iron, 0.107. When cooling 10 pounds of iron 1°F, one pound of water becomes 1°F warmer.

Latent heat of fusion is the amount of heat required to change a unit mass of a substance from a solid to a liquid state at the same temperature.

a

b

c

Figure 9.16 Ice-thrust features at the Iowa Great Lakes following the winter of 1969-1970. (a) Tree scour and overthrow. (b) Note size of boulders on rampart. (c) Grooved and striated nearshore lake bottom. Photographs by G. A. Miller, by permission.

One pound of ice at 32°F absorbs 144 BTU (80 cal/g) before it melts, and the temperatures of ice and water remain the same until all ice melts. Since there is no change in temperature to show the change in heat content, the heat is ''latent'' or ''hidden.'' When water freezes, the process reverses, and the same amount of heat is released.

Knowing these physical principles, let us examine a body of fresh water exposed to decreasing still-air temperatures. When the air becomes cooler than water, heat transfers from the upper layer of water to the air. The cooled water becomes more dense and sinks, establishing a vertical circulation which continues until the temperature is 4°C, when density is maximum. Further cooling reduces the density of the upper water layer, which floats, and the vertical circulation terminates. The heat transfer to the air continues, and the surface water layer freezes, forming lake ice, which then acts as an insulator reducing heat loss from the water to the atmosphere. The temperature of the ice layer is the same as air at the surface and 0°C at its bottom. A water layer beneath the ice has a temperature gradient downward from 0° to 4°C. Below this *thermocline,* water temperatures remain at 4°C.

This system is complicated by winds blowing across the lake; currents within the water body; convective air flow above the water, causing land-water air circulation; water depth; and the salt content of the water. These conditions may cause more or less rapid freezing or nonfreezing; or more rapid freezing along the shoreline.

When the surface water layer freezes, the shape of the ice crystals depends on the rate of freezing and the manner in which the water is exposed to low temperatures. Below the open surface of a lake, columnar crystals form at right angles to the water surface as *candle ice.*

Lake ice and shorelines

Upon complete freezing of the surface water layer, the expanded volume of ice causes thrust and overriding of the low shoreline. Where the shore resistance exceeds the shearing strength of the ice, the lake ice may buckle and become an overthrust, or be upheaved. The ridges thus formed are *ice ramparts* (fig. 9.14).

As the ice edge is thrust up the beach, it acts as a bulldozer blade and shoves beach sediments into ridges or *beach ramparts* (fig. 9.15). In shallow nearshore areas where freezing occurs to the bottom, ice entraps rock particles as large as boulders. With expansion, buckling, and thrusting shoreward, boulders are lifted and moved onshore. The ice scours, grooves, and striates the lake bottom in the nearshore area (fig. 9.16).

Examine ice activity on the Iowa Great Lakes, whose areas range from 0.2 to 9 square miles (fig. 9.16). They generally freeze over in early December, with ice breakup following in April. In 1969-1970, the average ice thickness increased from 10 inches in December to 20-25 inches in March (Miller, 1971). The heights of the ice ramparts on these lakes were 3.0 to 8.5 feet. The heights of the beach ramparts were 0.5 to 3.2 feet, and base widths were 1.0 to 18.3 feet.

Boulders more than a foot in diameter were moved as much as 20 to 23.5 feet and were thrust upward on ramparts as much as 2 to 5 feet. The stresses imposed by ice thrust moved concrete blocks weighing 170 pounds and a concrete slab, 5.0 x 16.0 x 0.5 feet, weighing about 6200 pounds. Lake ice, then, is able to modify shorelines.

The effect of lake ice on a strip of country is confined to the boundary of the basin. The effect of glacier ice is more widespread, since it occupies valleys, basins, and large parts of continents. Glacier ice confined in a valley is a *valley glacier*. Valley glaciers may descend to a basin or plain and coalesce laterally, forming a *piedmont glacier*. Both kinds of glacier conform to physiographic boundaries. An *ice cap*, or *ice sheet*, is not controlled physiographically and occupies large parts of continents.

A prerequisite for understanding glaciers and glaciation is an understanding of the mechanical properties of ice.

Mechanics of ice and ice flow

Like any crystalline substance subjected to stress, ice deforms elastically or plastically or may shear (Chapter 6). Let us consider a cylindrical material subjected in the laboratory to a confining pressure that acts over its entire surface. When an additional axial pressure is applied and the confining pressure is held constant, axial compression occurs. Axial pressure *(p),* or stress, is plotted against axial compression $\Delta L/L$, or strain, where L is cylinder length. The result is a stress-strain curve. Stress-strain-time relations are determined by measuring the deformation through time.

When a load is applied, ice deforms rapidly to a certain strain and then deforms more slowly with time (fig. 10.1). The slow deformation, *creep,* is characteristic of many solids. Creep enables ice to flow continuously under a given

10

Glaciation and Landscapes

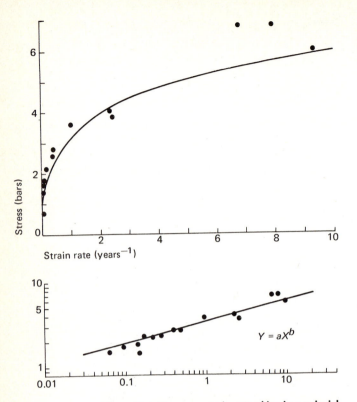

Figure 10.1 **The stress-strain rate of ice in a uniaxial compression test plotted arithmetically and logarithmically, giving the equation Y = aX^b. The strain rate is calculated as the ratio of deformation to time divided by compressed cylinder length.** From Glen (1952), by permission of Int. Glaciol. Soc.

stress. The stress-strain relation plots as a power law expressed as $Y = bX^n$ where Y is stress; X, strain rate; and b and n, constants (Glen, 1952, 1963). For ice at $-1.5°C$, $b = 1.62$ and $n = 4.1$ (Nye, 1952). The constants vary with amount and type of stress, rate of increase of stress, temperature, and bubbles and other impurities in the ice (Shumskii, 1964).

Glacier flow

Given this stress-strain relation, now known as the *Glen flow law,* the distribution of velocities of flow within glacier ice may be calculated (Nye, 1952), and vertical-velocity curves may be constructed (Chapter 3). Consider in principle the simplest case of flow, which is down a uniform plane slope (fig. 10.2). Lines of flow parallel the downslope bed gradient (a). Shear stress *(S)* on a layer at depth d measured perpendicular to surface *OX* is:

$$S = p\,g\,d\,\sin a$$

where p is a uniform density throughout thickness h and g is acceleration of gravity. Strain T is dV/dY where V is velocity.

The difference between the surface velocity *(Vs)* and velocity at depth d is expressed:

$$Vs - V = \frac{k}{n+1}\left(\sin^n a\right)\left(d^{n+1}\right)$$

where $k = (pg/b)^n$. The difference in velocity between the surface and bottom layers is determined by substituting Vb and h for V and d. The constants b and n are determined by shear test in the laboratory.

Once flow velocity is known, the discharge of ice through any cross section may be calculated (Nye, 1952). When these flow principles are applied (Meier, 1960), velocity-discharge relations may be constructed down a longitudinal profile of a valley glacier (fig. 10.3).

The principle of laminar flow has been oversimplified; laminar flow does not account for basal sliding or formation of transverse crevasses in a valley glacier [fig. 10.4(a), (b)]. Flow in glaciers is inclined obliquely downward from the surface in the accumulation zone and obliquely upward in the ablation zone (Sharp, 1954). The *accumulation zone* is that area where snow is converted to granular ice, *névé* or *firn,* at a more rapid rate than it is sublimated or melted. The *ablation zone* is that area of the glacier where losses, including calving, exceed additions. *Calving* is the spalling of masses of ice from the glacier's leading edge into bodies of water [fig. 10.4(c)].

Movement down the glacier may be either *compressive* or *extending flow* (Nye, 1952). If the forward velocity decreases down the glacier (fig. 10.3), ice is compressed. Flow lines incline obliquely upward toward the surface. If the velocity increases down the glacier, ice is extended, flow lines incline obliquely toward the base, and the upper layer is under tension, which causes transverse crevasses. In both cases thrust planes and shear faults form. Bottom and side drag also cause shear failure and crevasses.

Ice-mass budget and flow

The cause of glacier flow is accumulation of snow and its conversion to firn and ice in the zone of accumulation. Ice thickness is the force ($p\,g\,d$ in the shear-stress equation) causing flow from the zone of accumulation to the zone of ablation. The balance between additions and losses is the *glacier budget*. During the steady state the budget is zero, and the lateral margins and the terminus of the ice mass should be stationary. If accumulation dominates ablation, a positive budget will cause the ice margin to advance. If ablation dominates accumulation, a negative budget will cause the ice mass to shrink and the ice margins

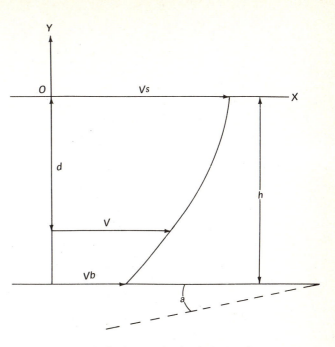

Figure 10.2 A vertical-velocity curve for glacier ice of thickness h on slope a: Vs, velocity of the surface layer; Vb, velocity of the basal layer; V, velocity of the layer at depth d. From Nye (1952), by permission of Int. Glaciol. Soc.

Figure 10.3 A longitudinal profile of the lower part of the Saskatchewan glacier, Alberta, Canada: T, thickness in feet; S, surface-layer velocity in feet per year; B, velocity of slide on the bed in feet per year; Q, discharge through a vertical rectangle of unit width in cubic feet per year. Velocity vectors to scale. Elevation and down-valley distance in feet. From Meier (1960).

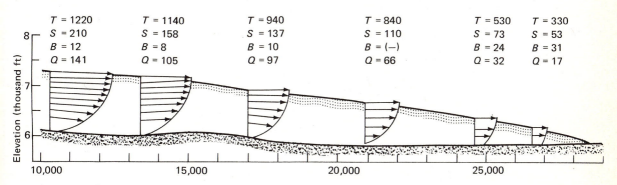

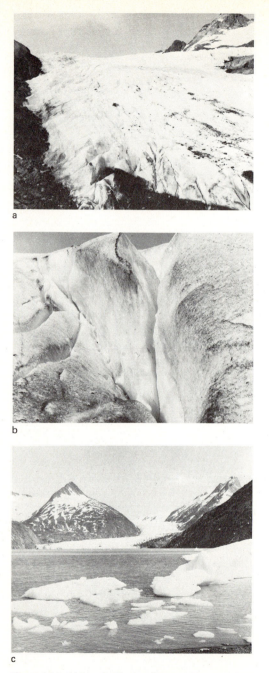

a

b

c

Figure 10.4 (a) Longitudinal and transverse crevasses with debris in, on, and around glacier ice. (b) Detail of transverse crevass and dirty ice. (c) Calving at terminus of ice.

to retreat. The budget is an adjustment to snowfall or climate, and the movement of glaciers is a sensitive response to climatic changes. The response in the glacier may not be directly related to the magnitude of the climatic change. A very slight fluctuation in climate may produce a great change in the extent of a glacier, and a major climatic change may cause only a slight change in a glacier's dimensions (Meier, 1965).

Some movements are not induced climatically but are caused by other dynamic conditions. Sudden rapid movements are *surges*. Black Rapids Glacier, locally known as Galloping Glacier, entered the valley of the Delta River from the west about 39 miles north of Paxson, Alaska. The historic marker along Richardson Highway reads: "Like an ice tidal wave this mile-wide column of ice moved to within one-half mile of the Richardson Highway during the height of its stage of advance in the winter of 1936. During this six months of great activity, the Galloping Glacier, 200 to 300 feet high, advanced a distance of four miles, giving an average daily advance of 115 feet. In February, 1937, the surging Black Rapids Glacier hit its maximum advance and began an energetic retreat. The glacier is still receding." This vivid description is matched by the raw nakedness of the surge debris 30 years later [fig. 10.5(a)]. The terminus of the receding glacier is many miles up valley and is just visible from Richardson Highway.

This surge was caused by heavy snowfall in the zone of accumulation during the years 1929-1932. About four to seven years were required for the dynamic adjustment within the glacier to cause the six-month surge of the terminus down valley (Flint, 1957). Contrast this surge velocity with the flow of the Saskatchewan Glacier, where surface velocities were 53 to 210 feet per year, and bottom rates were 8 to 31 feet per year (fig. 10.3).

In the continental ice sheet of Antarctica, the two types of flow are sheet flow and stream flow (Mellor, 1967). Where the thickness is greater than 6500 feet, ice slowly moves as a sheet, but

a

b

Figure 10.5 (a) Raw terminal and lateral moraine formed during the 1936-1937 surge of Black Rapids glacier in Alaska. The terminus of ice in 1967 is barely visible upvalley. (b) Rock debris in and on Worthington glacier, Alaska. Note the terminal moraine, size of boulders, morainal dammed lake, and calved ice.

near the coast, where the ice thins, ice streams 1 to 9 miles wide flow more rapidly within the sheets. Sheet flow is 365 feet per year, stream flow 1095 feet per year.

Let us examine the rates reconstructed from glacial moraines in the midwestern United States. The last glacier ice in Iowa extended as a lobe southward from the Minnesota state line 133 miles to the city of Des Moines. The maximum width of the lobe was 104 miles. Spruce forests were buried by the drift of the ice sheet, and the age of the wood has been determined by radiocarbon dating (Ruhe, 1969a). Near Scranton, Iowa, the average age is 13,865 years. Near Nevada, Iowa, the age is 14,200 years. The ice border was 12 miles southwest of Scranton and 20 miles southeast of Nevada. Near Colo, Iowa, the base of a peat bog *on* the drift lobe is 13,775 years old. Glacier ice thus buried forests, advanced to the terminal position, and uncovered an area 5 miles behind the terminal position during those times. Using burial date minus uncovering date and distance from burial site to terminal position, the rate of flow southeast of Nevada was 248 feet per year, southwest of Scranton, 704 feet per year.

In other areas of the Midwest the rates of advance, also based on radiocarbon dating, were 131 to 273 feet per year in Illinois (Kempton and Gross, 1971) and 41 to 481 feet per year in Ohio (Goldthwait, 1958). These values for the Midwest are comparable to the measured velocities of current glaciers.

Erosion, transport, and landforms

As demonstrated in the discussion of lake ice (Chapter 9), ice is able to shove things around and to pick them up and move them. This capability is magnified enormously in glacier ice. Pure ice has a Mohs hardness of 1.5 at $-5°$ C, so it can have little effect in abrading rocks except for the softest of materials. Erosion by glaciers is done by the

rock tools that the ice contains. As ice flows on its bed, it moves rock particles by shoving, scraping, and dragging, by plucking rocks, and by refreezing basal melt water around rock particles and plucking them. Like the abrasive particles on sand paper, partially enclosed tools abrade adjacent material.

Erosion features

Glacial-erosion features range from extremely small to large. As a tool held by the ice moves across bedrock, it may scratch, groove, and polish the overridden surface. *Striations* are fine, linear scratches; larger ones are *grooves*. The trace shows the local direction of ice movement, and the regional pattern indicates the direction of ice flow (Flint et al., 1959). *Crescent marks,* or *chatter marks,* are crescent-shaped gouges on rock caused by the sporadic impact of a tool carried by ice. They may be concave or convex, either with or against ice flow, but their axes of symmetry indicate approximate direction of movement.

Ice flow streamlines landforms, whose form is like the shape of the inverted bowl of a teaspoon. *Stoss-and-lee forms* are on bedrock with an upflow end that is generally smooth but striated and grooved. The downflow end may be jagged. The smooth end is formed by abrasion as ice streamlines around a local high bedrock. Downflow roughening is caused by quarrying due to frost wedging and plucking by moving ice, by plucking due to frictional drag, or by release of stress in the bedrock, causing breakup and plucking.

Drumlins are streamlined forms composed of unconsolidated debris deposited by glacier ice. Their type area is Ireland, (the term is derived from the Gaelic *druim,* "hill ridge"). Drumlins occur in fields or swarms, and their profiles from a distance resemble a school of whales breaking water [fig. 10.6(a)]. Their long axes indicate direction of ice movement (Flint et al., 1959), with the steeper end opposing flow.

Figure 10.6 (a) A drumlin field near Ballinamore, Ireland. (b) Cirques, hanging valleys, and U-shaped cross profile of valley in mountain glaciation. (c) Quarried headwall of cirque and cirque lake. (d) Low tide in fjord.

The origin of drumlins continues to be a matter of speculation. Two views are generally held: (1) they are formed by erosion of pre-existing glacial drift, or (2) they are depositional features (Gravenor, 1953). Both explanations require actively flowing ice. In the erosional theory, a first ice advance provides an irregular glacial till surface. After ice uncovers this surface, a second ice advance gouges and scours the surface into the streamlined forms of drumlins. In the depositional theory, glacier ice deposits a glacial till nucleus. With further ice flow, other layers of till are added, and the form is shaped and streamlined by the flowing ice. It may be that drumlins form both ways (Embleton and King, 1968).

Alpine sculpture is a scenic and spectacular kind of erosion caused by mountain glaciation. *Cirques* are scoop-shaped depressions at the heads of mountain valleys or on mountain slopes [fig. 10.6(b)]. Their form is dominantly concave. A longitudinal profile of a cirque is concave below a steep sheer headwall, and a transverse profile is concave between opposing steep, sheer sidewalls. The floor commonly is lower than a berm at the outlet, which impounds water and forms a *cirque lake* [fig. 10.6(c)].

Cirques are formed by a combination of processes. Melt water in a firn field or ice freezes in rock joints, wedging material from the rock mass. The material is transported down valley by a *cirque glacier*. On head- and sidewalls, frost wedging removes blocks from the walls, and they tumble as mass-waste products downslope [fig. 10.6(c)] or down firn or ice. A ridge of debris may accumulate at the end of the firn field or ice, forming a *protalus rampart*. The combination of freezing, thawing, and mass wasting are *nival processes,* or *nivation*.

Where adjacent cirques are closely spaced, ridges between back-to-back side- or headwalls are sharp and jagged *arêtes,* or *comb ridges*. Where three or more cirques have opposed headwalls, an intervening sheer, pyramidal mountain is a *horn*. The repeated occurrence of cirques, arêtes, and horns forms alpine sculpture.

Not all erosion by glaciers in mountains gives rugged landscapes. On highlands local icecaps create rolling slopes and subdued topography [fig. 10.6(d)]. Glaciated valleys have smooth transverse U-shaped cross sections. No satisfactory mechanical theory explains this profile (Scheidegger, 1970), although it has been attributed to frictional drag being minimal in a semicircular cross section (Flint, 1971).

Glacier ice does not *create* valleys. It occupies pre-existing stream valleys and in doing so deepens and widens the original features. Examine a *hanging tributary valley,* whose mouth may be many feet above the floor of the main valley [fig. 10.6(b)]. A greater volume of ice in the main valley eroded rock at a more rapid rate, leaving the tributary valley suspended at higher elevation.

Glaciers along coasts commonly descend below sea level, and their valleys later are inundated by sea water, creating *fjords*. Some of the valleys may have been glaciated on land while the sea was lower and then later submerged during a rise in sea level [fig. 10.6(e)]. Others may have been excavated by ice below sea level. Given the density differential between ice and sea water, ice can scour to a depth below sea level equal to about 90 percent of the thickness of the ice. Ice scour is common in fresh-water lakes [fig. 9.16(c)].

Many of these erosional features are not restricted to glaciation. Striations may form by a rock's scratching another in any process. Hanging valleys may result from differential stream erosion or more than one episode of stream erosion in the same valley. But in an area not currently under glacial conditions, an association of some of these features may indicate former glaciation.

Magnitude of erosion and transport

The magnitude of glacial erosion is difficult to quantify. In valley glaciation the amount of valley deepening and widening caused by prior stream activity must be separated from that caused by ice activity. It is also difficult to establish a datum for determining erosion by icecaps and ice sheets. It is generally believed that glacial erosion has been deep in mountain areas but slight in regions of low relief overrun by ice sheets (Flint, 1971).

In local areas, depth of scour may be exceptional. The rock floors beneath the Seneca, Cayuga, and Onondaga Lakes in New York are below sea level (Muller, 1965). At Ithaca a well near Lake Cayuga bottomed at a depth of 600 feet below sea level without entering bedrock. Tributary valleys hang well above the present lake levels. Yet on the adjacent upland, weathered bedrock is beneath glacial debris. Glacier ice locally scoured pre-existing valleys to great depths but had little erosive effect on uplands.

Evidence of erosion and transport is the debris deposited by glacier ice [fig. 10.5(b)]. Let us re-examine the Des Moines drift lobe in Iowa, which occupies an area of about 12,300 square miles (Ruhe, 1969a), where the average thickness of glacial deposits is 35 feet (Kay and Graham, 1943). The volume of debris is about 81.5 cubic miles. There *erratics,* rocks foreign to local bedrock, include Cretaceous shales from the Dakotas, granite from Minnesota, and native copper from northern Michigan. The calculated volume becomes quite thin when spread across the possible source region.

Regardless of these facts, a view is held that ice sheets deeply erode and carve great ellipsoidal basins (W.A. White, 1972). In North America the largest is the one with Hudson Bay as the master basin and the Great Lakes as lesser radiating basins down the ice sheet. A smaller system has eastern Lake Ontario as the master basin and the Finger Lakes of New York as lesser radiating basins. Large areas of Precambrian rocks correlate with the central core of the ice sheet, but down glacier a bounding peripheral region has Paleozoic and progressively younger rocks. Exposure of the Precambrian core area requires glacial erosion of hundreds to thousands of feet of these younger rocks.

If this is true in the central core area, the same principles must apply in the belts of progressively younger sedimentary rocks in the peripheral glaciated zone. Yet buried soils with all of their horizons (Chapter 2) are common features on bedrock, where they are buried by younger glacial debris. Buried soils are also common features separating one glacial deposit from an overlying glacial deposit throughout the region of the peripheral glaciated zone. How deep can glacial erosion be that fails to strip soils?

The transport capability of glacier ice is more easily quantified than effective erosion, since debris on, in, and around the ice can be measured (fig. 10.5). In areas formerly glaciated, the product of the deposit area and average thickness gives the volume transported.

The distance of transport is measured directly from an erratic to its source. A cobble of native copper in glacial till at Waterloo, Iowa, had to travel about 350 miles from its source in the Keeweenawan Peninsula of upper Michigan.

The competence of ice may also be measured directly in existing glaciers and in areas formerly glaciated. The size of transported material varies from clay to boulders, and the boulders can be enormous. In northeastern Iowa, boulders 10 to 25 feet in diameter were common, with some rocks as large as 35 to 50 feet in diameter (Alden and Leighton, 1917). The largest measured boulder was 50 × 40 × 11.5 feet (exposed part only). Most of these rocks were granite, and their nearest source was 185 miles to the north in Minnesota. In northeastern Iowa and in other places, these large erratics are aligned cross-country, forming *boulder trains.*

Glacial deposits and landforms

All rock material in transport in glacier ice, all deposits made by ice, and all deposits made by glacier melt water are *glacial drift. Glacial till,* unsorted and unstratified drift, may contain particles ranging in size from clay to boulders. *Stratified drift* has the properties of ordinary alluvium, and the mode of deposition of this kind of drift is the same as that of alluvium (Chapter 3, p. 60). Identification of stratified drift requires proof that glacier melt water is or was the transporting agent. In glaciated areas without current glacier ice, much ordinary alluvium is called stratified drift simply because it is in a glaciated region. Between the end members of glacial till and stratified drift there are many gradations from unsorted and unstratified to sorted and stratified sediment—classifying a material may be arbitrary.

Glacial till (there is a whole book on the subject) (Goldthwait, 1971) is deposited by basal lodging and ablation. Debris may lodge on the bed beneath the ice through frictional interference and shearing of ice above the basal load. Basal melting may simply deposit debris beneath the ice. The debris lodged beneath the ice is *basal till*, or *lodgment till*. *Ablation till* is deposited when ice melts and enclosed debris is lowered to a bed.

Boulders concentrated at the base of lodgment till form a *boulder pavement*. The long axes of these boulders or cobbles within the till may be directionally oriented. Measurement of the bearings of the axes is termed *till-fabric analysis*. Many cobbles are measured at a site, and a rose diagram is prepared showing the number or percentage of axes aligned in specific directions. Many sites are measured, and the diagrams are plotted on a map. A dominant direction emerges from the pattern that indicates the direction of ice movement in the area.

Glacial till commonly has inclusions of silt, sand, or gravel that are discontinuous lenses or beds. Deformed inclusions may be drag folds. Underlying buried soils may be dragged into an overlying till. Organic materials such as wood or peat may also be enclosed. Anything overridden by glacier ice may be incorporated in later deposited till.

Stratified drift deposited by running water has properties like alluvium and may be horizontally and cross-bedded. These relatively undisturbed structures form beyond the limit of the ice, and this stratified drift is *outwash*.

Stratified drift may be severely distorted with the angular juncture of beds, folds, and faults. The structures represent *ice-contact deposits*, and distortion occurs when ice support is lost because of melting causing sag, slump, collapse, and flow.

The activities of glacier ice, melt water, and ice-contact phenomena organize glacial drift into various constructional landforms and landscapes.

Till landscapes

Glacial till is the building material of specific landforms and landscapes. *Moraine* is originally a French term applied to ridges of rock and earth debris around the glaciers in the Alps. An *end moraine* is a ridgelike accumulation of till with associated stratified drift built along the margin of a glacier (fig. 10.5). Many ridges are parallel or subparallel and extend cross-country in an arcuate pattern forming an *end-moraine system* or *belt*. When mapped areally, the ridges are seen to have the pattern of a lobe, and the outermost ridge of the pattern is the *terminal moraine*.

End moraines of continental ice sheets form large lobate patterns. In the United States they extend from the Rocky Mountains to the Appalachian Mountains. In the Midwest specific drift lobes are oriented along prior topographic lows (fig. 10.7). The Des Moines lobe extends southward from the Red River lowland along the east border of the Dakotas, southeastward through south-central Minnesota, and southward into central Iowa along bedrock lows. Oriented end moraines delineate the Superior lobe around Lake Superior basin and the Green Bay lobe, which extends from Green Bay of the Lake Michigan basin into southern Wisconsin. The Lake Michigan lobe occupies large areas in Michigan and Wisconsin and extends southward

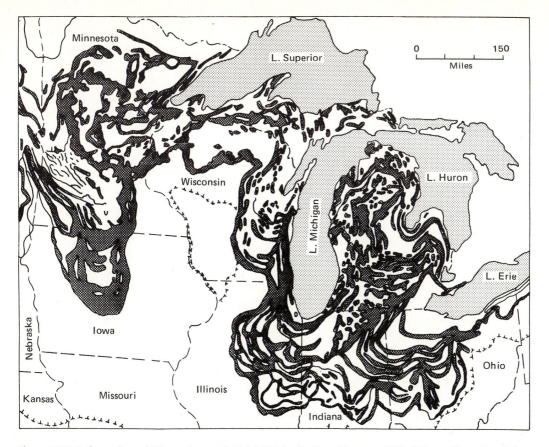

Figure 10.7 End moraines of Wisconsin-age glacial drift lobes in the midwestern United States. The end-moraine pattern is in heavy black. The hachured line is the drift border. From Flint et al. (1959), by permission of Geol. Soc. Am.

into southern Illinois and Indiana. The Saginaw lobe is aligned around Saginaw Bay of Lake Huron basin and covers a large part of lower Michigan. The Erie lobe is distinctly marked by moraines that extend southwesterward into central Indiana. According to radiocarbon dating these end-moraine systems were constructed 20,000 to 10,000 years ago (Bryson et al., 1969). The oldest moraine is the outermost system in northwestern Iowa, and this moraine also extends from Illinois across Indiana into Ohio. The youngest moraine crosses northern Minnesota

and borders Lakes Superior and Huron. Note the lateral mergence of end moraines as an *interlobate moraine* between adjacent lobes, as in southeastern Wisconsin (fig. 10.7).

A unique pattern of closely spaced ridges of very low relief may occur within an end-moraine belt. On the Des Moines lobe in Iowa and elsewhere, there are as many as 15 to 20 parallel ridges in a mile, and the maximum local relief is only 5 to 10 feet. These features are *minor moraines* (Gwynne, 1942, 1951), or *washboard moraines* (Lawrence and Elson, 1953).

In valley glaciation, a *lateral moraine* forms along the sides of the glacier between the ice and the valley wall [fig. 10.5(b)]. Lateral moraines from adjacent tributary valley glaciers merge as a *medial moraine*, which separates two ice streams in a main valley.

All of these moraines that are aligned as ridges mark the edge of active glacier ice during some period. Two general terms, push and dump, apply to modes of origin of the ridges. Ice may push or shove material into a ridge in front of it much the way the blade of a bulldozer does. *Push moraine* is the term applied to ridges so formed. Some of the minor moraines are believed to have been formed by the push mechanism.

Morainal ridges also form by material being dumped down the steep slopes of an ice terminus. Within the active ice, material is transported along shear planes to the end slopes. Material carried on the ice is also dumped in this manner. With ice wastage during a stagnant ice phase, material in and on the ice in the terminal zone is simply lowered to the ground.

Ground moraine is distinguished from end moraine by lack of alignment of the topographic highs and lows. The swell-and-swale pattern is random. Ground moraine is usually behind its associated end moraine, and this pairing is recognized in the *morphostratigraphic unit*, which is composed of a body of till and stratified drift and is recognized by the surface form that it displays (Frye and Willman, 1960). Each unit consists of the end moraine and ground moraine and the continuation of the body in the subsurface where it can be recognized. In Illinois, where the unit was first established, each end and ground moraine is delineated from the outermost moraine of the Lake Michigan lobe northeastward to the lake shoreline. Each younger end and ground moraine is superimposed on the next, older unit. Consequently the older may be identified and traced in drilling through the younger unit.

The construction of end and ground moraine creates new landscape, and because of the unequal distribution of load during ice transport, the landforms have initial topographic highs and lows. Both moraines have closed depressions that contain ponds, lakes, and bogs. The constructed land surface is poorly drained. On the Des Moines lobe in Iowa, poorly drained soils occupy major portions of the drift landscape (fig. 2.14). In Story County, in the southern part of the lobe (fig. 10.7), about 30 percent of 363,000 acres contains poorly drained soils in upland depressions (Meldrum et al., 1941). In Kossuth County, near the Minnesota state line, about 53 percent of 629,000 acres contains poorly drained upland soils (Benton et al., 1925).

The construction of poorly drained landscape may be on a large scale. Lakes may be impounded behind an end moraine and in front of the glacier ice [fig. 10.5(b)], creating a *morainal dam lake*. In Illinois, Lakes Ottawa, Pontiac, Watseka, and Wauponsee are behind the Farm Ridge, Cropsey, Chatsworth, and Marseilles moraines respectively (Willman and Frye, 1970). Glacial Lake Agassiz was impounded behind the Big Stone moraine of the Des Moines lobe in Minnesota and South Dakota. Impounding also occurred by ice blockage of drainage lines, forming *ice-dammed lakes* in front of the ice and now beyond the terminal moraine.

Drift-lobe construction also creates new drainage patterns (fig. 10.7). They merge northward at the Packerton interlobate moraine. Drainage nets have formed on the end moraines (fig. 10.8), but intervening ground moraines and lake plains have not had similar development. The end moraines provide a topographic and structural control for the drainage pattern.

Glacial drift-lobe construction creates a widespread veneer of surficial deposits that cover pre-existing drifts or landscapes. Some lobes, such as the Des Moines lobe in Iowa, have little lithologic difference from one end- and

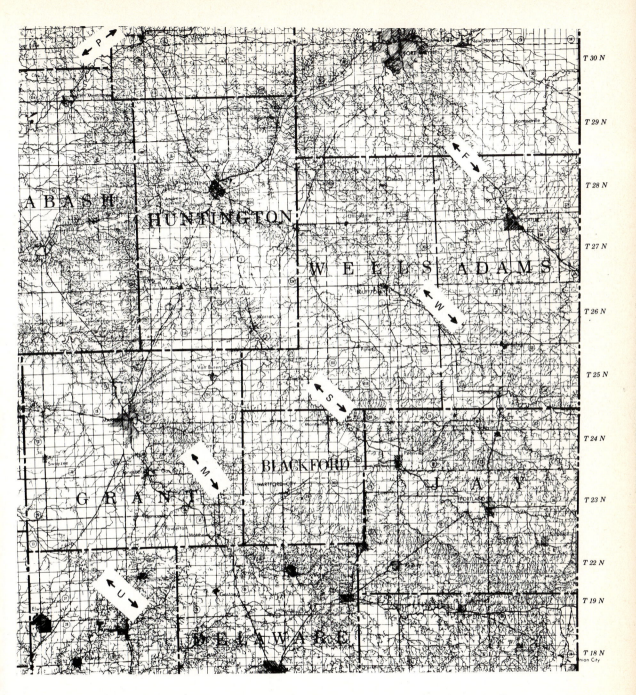

Figure 10.8 The relation of drainage pattern to end and ground moraines and lake plains in northeastern Indiana. Concentric arcuate moraines are Fort Wayne (F), Wabash (W), Salamonie (S), Mississinewa (M), and Union City (U); they merge northward at the Packerton interlobate moraine (P). From Engr. Exper. Sta. Staff, Purdue Univ. (1966), by permission.

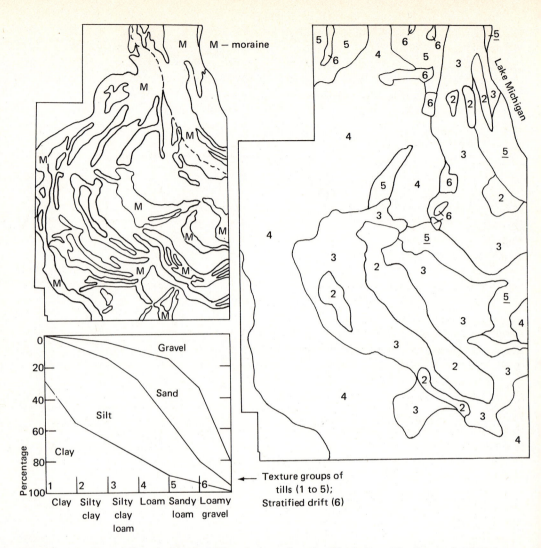

Figure 10.9 Textures of tills and other glacial deposits in northeastern Illinois. From Wascher et al. (1960), by permission of Ill. Agric. Exp. Sta.

ground-moraine unit to another. On the Lake Michigan lobe in northeastern Illinois (fig. 10.7), concentric bands of till of different textures conform to concentrically banded end and ground moraines (fig. 10.9). Soils inherit these parent materials (Wascher et al., 1960).

Stratified-drift landforms

Stratified drift is bedded and sorted, and running water is active in transporting and depositing the sediment. Erratics are as common in stratified drift as in glacial till.

Stratified drift confined to a valley is a *valley train*. It may be a terrace along the valley walls or bottom land. Stratified drift spread out across the land is an *outwash plain*, which may be pocked with closed depressions, or *kettles*. Kettles are usually formed by the burial of a block of ice, whose subsequent melting causes collapse of stratified sediments, which forms the depression.

Isolated hills of stratified drift are *kames*. The sediment is deposited in depressions or crevasses in glacier ice. Upon melting of the ice the sediment is left in the form of a mound or ridge. The ridges generally are termed *crevasse fillings*. In valley glaciers, stratified drift deposited between the valley wall and the lateral margin of the ice is a *kame terrace*. When the ice disappears the landform is a terrace bordering the valley.

A serpentine ridge of stratified drift is an *esker*. Eskers form as ice-channel fillings confined by walls of ice, and after the ice melts, they remain as ridges. Most eskers probably form in tunnels at the base of glacier ice.

Kames, kettles, crevasse fillings, and eskers represent a condition of stagnant ice and also represent the waning phase of glaciation. They are ice-contact features that become landforms by the wasting away of glacier ice. Collapse structures are common in all of them. If many of these landforms are present in a given area, they typify *collapse drift*.

Glacial stages

A regional distribution of glacial till and stratified drift is a *drift sheet*, which may occupy sizable portions of a continent (Flint et al., 1945, 1959). In some places earlier drift sheets were constructed far beyond the outer limits of later drift sheets. In the Missouri River basin, the limit of the older drift sheets in Nebraska, Kansas, and Missouri is far south of the younger drift sheet marked by end moraines in Iowa (fig. 10.7). Through Illinois, Indiana, and Ohio the same relation holds true. The separation of drift sheets is based on differences in weathering, soils, and erosional modification of the older and younger drifts.

In other places a younger drift is deposited on an older drift. They are separated stratigraphically by employing the principles of *Pleistocene stratigraphy*. The classic separation of glacial stages in the United States was done in the Midwest and because of its historic priority is the standard for North America.

Separation of drift sheets

Older drifts are altered more severely by erosion, weathering, and soil formation than younger drifts. In Iowa (fig. 10.7) the youngest drift surface, which is 13,000 to 14,000 years old, has little integrated drainage (fig. 5.2). The next younger drift surface, 14,000 to 20,000 years old, has a denser drainage net. Still older surfaces have more closely cropped drainage (fig. 5.2). The differences in drainage densities show the modification of each surface by erosion since it was uncovered by glacier ice, and the values are indicators of relative age.

The youngest surface maintains the constructional features of end moraine, washboard moraine, and poorly drained ground moraine. The next younger surface has only small remnants of end moraine and little poorly drained ground moraine, and most hillslopes descend to the integrated drainage net. Constructional features have been modified by erosion. Differences in surface drainage are a measure of postglacial erosion and are a criterion for separating adjacent drift sheets.

Another method of differentiation is measuring weathering characteristics. The depth of leaching of carbonates has been used for many years. In Iowa the mean depth of leaching on a younger drift surface is 30 inches; on the next younger surface, 66 inches. Time increments of 1.0 and 2.2 were assigned for the duration of weathering of the two drifts. Assuming an age of 25,000 years for the younger, the older is 55,000 years old (Kay, 1931). In Indiana carbonates are leached to a mean depth of about 4.5 feet on a drift somewhat older than the younger drift of Iowa. Using the 25,000-year base in Iowa, the older drift in Indiana is 45,000 years old (Thornbury, 1940). These estimates, which assume a linear relationship between leaching depth and time, were made prior to radiocarbon dating. The relative values were not unreasonable because C^{14} dates approximately halved the estimates. The datum drift in Iowa is 13,000 to 14,000 C^{14} years old; the older drift in Indiana, 20,000 to 22,000 C^{14} years old.

This may be coincidental, however, since depth of leaching is more complicated than a straight-line function with time. Among the factors affecting weathering (Chapter 2) are texture, permeability, initial carbonate content, rainfall, depth of wetting-front penetration, precipitation of secondary carbonate, temperature, transpiration by plants through their root mat extracting soil moisture and causing precipitation of secondary carbonate, and depth to a saturated zone of ground water. Of these factors carbonate content may be most important (Merritt and Muller, 1959), and combinations of factors must be manipulated to adjust for differences in carbonate content (Dreimanis, 1959).

This technique was applied to closely related drifts in Michigan (B. L. Allen and Whiteside, 1954), where mean depths of leaching of two tills

are significantly different statistically: 20.2 ± 1.2 and 29.8 ± 1.5 inches. Respective carbonate contents are also significantly different: 31.5 ± 1.5 and 22.9 ± 1.0 percent. The amount of carbonates leached from each drift is approximated by the product of mean carbonate content and mean depth of leaching, or $20 \times 31 = 620$ and $30 \times 23 = 690$ for the younger and older drifts respectively. Assigning 1.0 to the younger, the older is 1.1; if the younger is 11,000 years old, the older is 12,100 years old. Now the radiocarbon isochrones are checked in the immediate area (Bryson et al., 1969), which are 11,000 and 12,000 years old. With careful measurement, depth of leaching may be useful for drift separation (Dreimanis, 1959) regardless of criticism of the method (Flint, 1949).

Since depth of leaching is a soil property, soils are useful in separating drift surfaces. The Clarion-Nicollet-Webster soil association is restricted to the Des Moines drift lobe in Iowa, but other soils are on all surrounding landscapes.

Soil stratigraphy

Drift sheets may be separated stratigraphically by glacial or nonglacial features. If the only means of separation is difference in composition, the tills may represent only lithologic phases of one glaciation. If soils, weathering profiles, peat, muck, forest beds, sediments containing fauna or flora, alluvium, or swamp, lake, or marine deposits are between tills, glacial deposits are separated by nonglacial features. One may then separate deposits into major divisions of glacial stages or stades or into subdivisions of glacial substages or substades. The intervening features may represent interglacial stages (interstades) or intraglacial substages (intrastades).

Soils and weathering profiles are primary separators. A soil on one glacial drift that is buried by another drift is a *buried soil* or paleosol. A paleosol is a soil that formed or began forming on a landscape in the past (Ruhe, 1965a). The study of paleosols is *paleopedology*. Yaalon's book (1971) on the subject is recommended. A buried soil between tills shows that during an interval, glacier ice was not present, and soil formed on a landscape. Hence the buried soil is an intraglacial or interglacial feature.

Buried soils are recognized formally in soil stratigraphy. A *soil-stratigraphic unit* is a soil with features and stratigraphic relations that permit its consistent recognition and mapping as a stratigraphic unit and that is distinct from rock-stratigraphic and pedologic units. The soil-stratigraphic unit forms by the weathering of an underlying rock-stratigraphic unit and may extend cross-country on many underlying units. The soil-stratigraphic unit is defined at a type locality by its features and relations to overlying and underlying rock-stratigraphic units and cross-country by its lateral variations and stratigraphic relations. The unit may parallel or transgress time (Richmond and Frye, 1957; American Commission on Stratigraphic Nomenclature, 1961). Pedologic units are the various categories in formal soil classification (Chapter 2).

To determine whether a soil-stratigraphic unit has interglacial or intraglacial rank, the principle of uniformitarianism is applied. The properties of the soils and weathering profiles on the youngest, uncovered glacial drift are used as a datum for comparison with buried features. In the leaching of carbonate in Iowa, average depth is 2.5 feet on the youngest drift (Kay, 1931). A next younger drift is leached 12 feet where it is buried by the youngest drift. A third drift is leached 30 feet where it is buried by the second drift, and a fourth drift is leached 20 feet where it is buried by the third drift. Setting the surface leaching depth at 1.0, the buried leaching intervals successively downward are 4.8, 12.0, and 8.0. The lower intervals have greater magnitude than the surface one, which represents all of postglacial time. Consequently, the lower buried leaching intervals are given the major rank of interglacial stages.

Other properties of soils, such as the amount of clay in the solum, the weathering status of minerals, and the thickness of solum, are also used for comparison purposes. These three criteria were used to evaluate *gumbotil*, in part a weathered product of till (Kay, 1931). Gumbotil is an upper zone in a weathering profile (Leighton and MacClintock, 1930). If there is accretion in depressions on a till surface, the weathered material is *accretion-gley* (Frye et al., 1960).

If gumbotil and accretion-gley are studied and analyzed pedologically, they are found to be various kinds of paleosols (Simonson, 1941, 1954; Ruhe, 1956b, 1965a, 1969a), and their thicknesses are generally much greater than the solum of soils (Chapter 2) on the youngest drift surface. In Iowa, corresponding to the vertical sequence described for depth of leaching, respective solum thicknesses are less than 3, 5, 12, and 8 feet. These are mean thicknesses of the ground soil and successively lower buried soils. In addition, the paleosols have greater clay contents and lesser amounts of weatherable minerals (Ruhe, 1969a). This contrast in properties also shows that the paleosols have endured greater weathering than the ground soil and are interglacial representatives.

To establish the intraglacial rank of a paleosol, the magnitude of its properties must be less than those of the interglacial soil and is usually less than those of the youngest postglacial soil. All of these principles are used to separate and classify drift sheets. Some others are needed for classifying mountain glaciation.

Criteria for glacial stages in mountain glaciation

In mountain terrain, relief may change greatly in relatively short distances, and erosion of drifts may be more severe than in flat terrain. Since climatic zones change with elevation, the weathering of drift varies. Because of their confinement in valleys one drift may be inset below another drift much like low and high terraces along a valley wall.

Let us consider the older and younger drifts in the Wind River Mountains of Wyoming (Richmond, 1965). Three older moraines descend valleys to respective valley trains that cap rock terraces 200, 100, and 80 feet above the major drainageway. These end moraines are large, have smooth slopes, are broadly breached, and do not contain lakes. The lateral moraines are notched by tributary streams. Boulders are numerous on the surface of the moraines, are broken or exfoliated, and are stained to a depth of about a centimeter. The moraines extend down valley a maximum distance of 20 to 25 miles.

Three younger moraines rest on, breach, or lie just up valley from the older moraines. Breaching is in the form of narrow tongues. The younger moraines are steep and irregular and are usually smaller than the older features. The younger end moraines are narrowly breached by erosion, and the lateral moraines are little modified. Lakes are common. Boulders, which are very numerous on the surface, are fresh, and few of them are fractured.

Soils on older deposits have redder, thicker, more clayey B horizons and thicker Cca horizons than those on younger deposits. Where younger drift overlies a paleosol on older drift, the buried soil has a developed solum.

Age distinctions are made on the basis of geometry, morphology, general aspects of weathering and erosion, and degree of soil development. Where gross lithologic and topographic features prevent application of these techniques, other methods, such as surface-boulder frequency and boulder-weathering percentages, must be applied (Sharp, 1969). Sites are selected that are comparable in topographic setting and lithologic mix. Closely spaced lateral moraines on the same side of a valley are used, and measurements are made in many places. Surface stones larger than a specific diameter are counted within strips of a standard length and width.

Within a boulder population of a specific rock type, any boulder retaining a smooth abrasion surface is considered unweathered, all others weathered. A count frequency is expressed:

$$A\ (\%) = \frac{A}{A + W} \times 100$$

where A is abraded and W is weathered. A count is also made of boulders that are pitted or fretted or show other evidence of granular disintegration. This count frequency is expressed:

$$F\ (\%) = \frac{F}{F + W} \times 100$$

where F is fretting. These techniques approximately quantify the weathering of surface boulders exposed on one drift compared to the weathering of boulders on another drift.

By associating many of the features of mountain glaciation, one can trace given stages cross-country many miles along the axes of mountain ranges and from range to range. In the Rocky Mountains specific stages have been traced from the Canadian border almost to Mexico (Richmond, 1965). The average altitude of moraines of a given stage are near 4000 feet in Glacier National Park and near 10,000 feet on Sierra Blanca in southern New Mexico. A north-to-south climatic stratification is evident. In fact, the presence of moraines on mountains and the occurrence of drift sheets in the midcontinent, where glacier ice no longer exists, is direct evidence of different climates and environments in the past.

Many earth features show that the environment in the past differed from that of the present at a given place. *Environment* is the aggregate of surrounding things, conditions, or influences that affect the existence or development of something. Glacial deposits in the midcontinental United States (Chapter 10) demonstrate a glacial environment in the past where there is a nonglacial environment today. Large dry-lake basins in the now arid Southwest (Chapter 9) indicate past differences. Widespread loess deposits in the central United States (Chapter 8) show a change of environment from past to present. Most changes are not instantaneous but occur in an orderly fashion through time. Through time, environment and change are the interaction and impact of climate and biota on landscapes and materials.

Impact of climate

There are three dominant patterns—climate, vegetation, and soils—on the earth, and their boundaries are almost coincident because climate is the principal cause of the other two (Blumenstock and Thornthwaite, 1941). The distribution of vegetation, soils, and land features depends on precipitation, temperature, evaporation, sunshine, and cloud cover. Of these climatic parameters, the first three are the most important and combine as an index of "effective precipitation" because evaporation generally increases with an increase in temperature.

Many classifications of climate use a precipita-

11

Environment and Change

tion effectiveness (PE) index, and one by Thornthwaite (1931, 1933), based on an analysis of climatic records and vegetation distribution, is expressed on a monthly basis as:

$$PE = \left(\frac{P}{T - 10} \right)^{\frac{10}{9}}$$

where P is precipitation in inches and T is temperature in °F. An index is calculated for each month, and the sum of the monthly indices is the annual index. Greater weight is given to rainfall during cold months, when evaporation is lower, than during hot months, when it is higher. Climatic moisture regions are separated by PE values: superhumid (wet), greater than 128; humid, 64 to 128; subhumid, 32 to 64; semiarid, 16 to 32; and arid, less than 16.

A temperature efficiency (TE) index of Thornthwaite is based on temperatures above freezing, which are only the temperatures beneficial to plant growth. The TE index is calculated by subtracting 32°F from each mean monthly temperature and summing for an annual index. This sum when divided by four gives values like the PE index and separates climatic temperature regions: macrothermal, greater than 128; mesothermal, 64 to 128; microthermal, 32 to 64; taiga, 16 to 32; tundra, 0 to 16; and perpetual snow and ice, less than 0. (Taiga and tundra really refer to vegetation, however, since taiga is coniferous forest and tundra includes mosses, lichens, and dwarf trees and shrubs of subarctic regions.)

If regional indices are plotted on a map, climatic belts, vegetation zones, and soil associations

Figure 11.1 Climatic zones based on the PE index of Thornthwaite (1941). Vegetation zones from Barnes (1948). Soils from Kellogg (1941), adjusted to N.C.R. - 3 Tech. Comm. Soil Survey (1960) and U.S. Department of Agriculture, Soil Conservation Service (1969). From Ruhe (1970), by permission of University Press of Kansas, Lawrence.

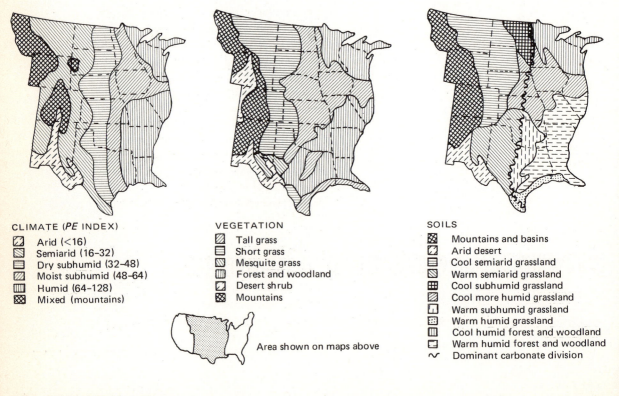

CLIMATE (*PE* INDEX)

- Arid (<16)
- Semiarid (16–32)
- Dry subhumid (32–48)
- Moist subhumid (48–64)
- Humid (64–128)
- Mixed (mountains)

VEGETATION

- Tall grass
- Short grass
- Mesquite grass
- Forest and woodland
- Desert shrub
- Mountains

Area shown on maps above

SOILS

- Mountains and basins
- Arid desert
- Cool semiarid grassland
- Warm semiarid grassland
- Cool subhumid grassland
- Cool more humid grassland
- Warm subhumid grassland
- Warm humid grassland
- Cool humid forest and woodland
- Warm humid forest and woodland
- ∿ Dominant carbonate division

are related. In the central United States climatic zones based on the *PE* index trend north and south, but one prominent projection of the moist subhumid zone extends east and west across Iowa and Illinois (fig. 11.1). Vegetation zones correspond well to the climatic belts. Short-grass prairie is a feature of the semiarid zone, tall-grass prairie, of the subhumid zone, including the projection across Iowa and Illinois, which is known as the *Prairie Peninsula*. Note the correspondence between humid climatic zones and forest and woodland.

Soil patterns generally correlate with the climatic and vegetation zones (fig. 11.1), but the factor of effective temperature must be considered. Superimposed on the north-south trending zones is a latitudinal stratification of temperature from cool in the north to warm in the south.

This threefold system can be expanded to a semiglobal scale (fig. 11.2). According to the *PE* index, the climatic zones are arid, semiarid, subhumid, humid, and wet; the correlative vegetation zones are respectively, desert grasses and shrubs, steppe, grassland, forest, and rain forest. (*Steppe* is an essentially treeless prairie.) The soil zones in order are Sierozems and Desert soils, Chestnut and Brown soils, and Chernozems and Brunizems, the last two conforming to dry subhumid and moist subhumid grasslands respectively.

In humid-forest and rain-forest regions, effective temperature is a dominant climatic factor. Where the *TE* index is 32 to 48, Podzols are the dominant soils. Gray-Brown Podzolic soils conform to a *TE* index of 48 to 64, Red-Yellow Podzolic soils to an index of 64 to 128, and Lateritic soils to an index greater than 128. This arrangement is a latitudinal zoning, with increasing temperatures from northern temperate to equatorial regions. In the other direction, effective temperatures are below freezing (less than 32°F), evaporation is low, and effective precipitation is not a dominant factor.

The combination of climate, vegetation, and landscape features is recognized in the *mor-*

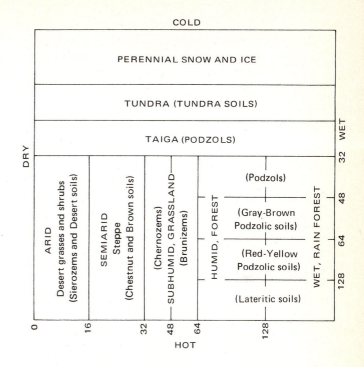

Figure 11.2 The relation of climate, vegetation, and soils from cold to hot and dry to wet regions. PE index along horizontal scale; TE index along vertical scale. From Blumenstock and Thornthwaite (1941).

phogenetic region (Peltier, 1950). Under certain climatic conditions specific geomorphic processes dominate, resulting in certain landscape characteristics that differ from those formed under other climatic conditions in other regions. Various intensities of glacial erosion, weathering, running water, mass movement, and wind are used to define nine morphogenetic regions.

On weathering and soils

Climatic forces, particularly precipitation and temperature, directly and indirectly influence kinds of weathering (Chapter 2). In dry, cold regions, physical weathering is more important than chemical or biological weathering. Conditions are optimum for heating, cooling, and frost action. In moist, hot regions, conditions favor chemical and biological weathering. The greater the effective precipitation and the effective temperature, the greater are the amount and degree of chemical and biological weathering. With a decrease in precipitation and/or temperature, physical weathering becomes more important.

The amount of effective precipitation directly controls the severity of leaching of material where downward water movement is uninhibited. A regional climatic boundary extends from the Canadian border to the Gulf of Mexico and separates a zone to the east, where precipitation exceeds evaporation, from a zone to the west, where evaporation dominates (fig. 11.1). The boundary between dry and moist subhumid regions separates *Pedalfers* (humid soils) to the east from *Pedocals* (arid soils) to the west (Marbut, 1935). Pedalfers are leached of carbonates but have accumulations of aluminum and iron. Pedocals have accumulations of carbonates in all or part of their soil profiles. With increased leaching, increased effective precipitation causes increased acidity and formation and accumulation of clay minerals.

An increase in effective temperature causes an increase in chemical weathering where effective precipitation controls leaching. The severity of weathering increases from Podzolic soils and Gray-Brown Podzolic soils to Red-Yellow Podzolic soils and to Lateritic soils. Through this sequence, silicate minerals become more depleted and clay-size minerals increase toward more kaolinite and sesquioxides (Chapter 2).

The most direct effect of the climate on soils is the formation of large soil structures known as *patterned ground,* which includes somewhat symmetrical circles, polygons, nets, steps, and stripes. These forms are characteristic of, but not necessarily restricted to, intensive frost action. Patterned ground is classified thus (Washburn, 1956): *sorted circles, nets, and polygons* have circular, intermediate, and polygonal ground patterns respectively with a border of stones surrounding finer material (fig. 11.3). *Nonsorted*

circles, nets, and polygons lack the stone border. A kind of nonsorted net is an *earth hummock* (fig. 11.4). *Sorted steps* are steplike forms with a downslope border of stones embanking finer material upslope. *Unsorted steps* have a downslope border of vegetation with barren ground upslope. *Sorted stripes* have parallel lines of stones with intervening strips of finer material on slopes. *Nonsorted stripes* have parallel lines of vegetation with intervening strips of relatively barren ground.

The origin of these structures is not well understood, and various explanations are given (Washburn, 1956). Frost wedging is capable of creating patterned ground. In Alaska the climate is latitudinally stratified from a mean annual temperature of +4°C along the southern Pacific coast to −12°C along the northern Arctic coast (Péwé, 1965). The 0°C isotherm parallels the southern coast slightly inland. A northern zone, where mean annual temperatures are −6°C to −12°C, has continuous permafrost, active ice wedges, and widespread patterned ground. Active ice wedges occupy the cracks, forming the mesh of the pattern (Péwé, 1966). A central zone, where mean annual temperatures are 0°C to −6°C, has weakly active to inactive ice wedges, discontinuous permafrost, and patterned ground. In this central Alaskan region, troughs bounding the polygons no longer have ice wedges but probably originated when mean annual temperatures were 3°C colder (Péwé et al., 1969). The southern zone, where mean annual temperatures are above freezing, generally lacks ice wedges.

In cold regions, patterns may form by expansion due to freezing, with ejection of stones from fine earth during repeated freezing and thawing. They may also form by contraction due to low temperatures (Lachenbruch, 1962). As temperatures decrease below freezing, ice contracts, and frozen ground with a large ice content develops frost cracks of tensional origin. Ice wedges form in the cracks of the polygonal pattern. With melting of the ice wedges, sediment washes or falls

a

b

c

into the voids, creating ice-wedge pseudomorphs or "fossil" ice wedges (Péwé et al., 1969). This introduction of more solids along the polygon's boundaries causes the surface of the polygon to dome or bow upward upon expansion during repeated freezing and thawing (fig. 11.3).

Patterned ground also forms by processes other than cold-region ones. Drying causes tensional-desiccation cracks, which form polygonal patterns—some of the polygons in playa sediments in the southwestern United States are as large as football fields (Willden and Mabey, 1961). Bounding cracks extend many feet downward (Ruhe, 1967). Sediment washes or falls into these cracks, and upon its wetting and expansion, the sides of the polygon abut the void fillings, which causes the surface of the polygon to bow upward, forming a mound. In these mounds heaving caused by repeated wetting and drying creates *gilgai,* severely distorted structures. The soil tends to overturn in a self-mulching process in these soils, which are known as *Grumosols* (Templin et al., 1956; Kunze and Templin, 1956). Gilgai in Australia occur not only as nonsorted nets but also as nonsorted stripes where parallel cracking occurs in the landscape (Stace et al., 1968). There gilgai form under 6 to 60 inches of annual rainfall, but regular or intermittent drought is needed so that the

Figure 11.3 (a) A sorted polygon with basalt pebbles bordering a fine-earth matrix near Carey, Idaho. See Hugie and Passey (1964). (b) A sand wedge in Tama County, Iowa. Scale is in feet. (c) The domed cross section of polygons buried beneath loess in Tama County, Iowa. From Ruhe (1969a), by permission. © Iowa State University Press, Ames.

a

b

Figure 11.4 (a) Earth hummocks on a mountain slope above Palmer, Alaska. (b) Nonsorted nets along Dinali Road near Paxson, Alaska.

soils can dry several feet deep during the high summer temperatures.

Intensively mounded ground in sorted and nonsorted nets is caused by biological activity in the central valley of California (Arkley and Brown, 1954). Fine earth is heaped up by pocket gophers in their burrowing and nest building. In some places gravels mantle the intermound areas between the *mima mounds* and give a sorted-net appearance to the field. Termite mounds commonly occupy the landscape in tropical areas, and near Stanleyville in now Zaire some forms, called *termiterre*, are many meters in diameter and many meters high. *Pimple mounds* are clusters of mounds constructed by crayfish in wet lowlands.

Toppling of trees, *tree throw,* may also form patterned ground. As a tree falls, mounds and pits form, and on beaches of the Champlain Sea near Plattsburgh, New York, the pattern looks like sorted circles (Denny and Goodlett, 1968). Trees also produce soil tongues or pendants, which extend downward around a tap root. After removal of the tree, the soil tongue may look like a "fossil" ice wedge (Yehle, 1954).

Since patterned ground forms in many ways, one must be careful in interpreting the paleoclimatologic significance of these features. Some of the features are difficult to explain by any process. For example, how does platy limestone shear from flat-lying beds and become arranged in vertical, imbricate festoons that have a polygonal plan on a flat-ground surface?

On geomorphic processes

The most obvious impact of climate on geomorphic processes is glaciation (Chapter 10), but other processes are also dependent on climate. The effectiveness of running water, mass movement, and wind changes as *PE* and *TE* vary (Blumenstock and Thornthwaite, 1941). When *PE* is relatively high, erosion by running water becomes less effective as temperature decreases because the length of the freezing period increases. Mass movement (Chapter 6) generally requires a relatively high *PE* but may be effective throughout the entire *TE* range. At the warm end of the range, slumping is dominant, and at the cold end solifluction is significant. When *PE* is relatively high, wind effects are unimportant, as the ground is protected by vegetation.

If *TE* is relatively constant, the effects of running water, mass movement, and wind vary with *PE*. Erosion by running water generally decreases as precipitation lessens. Mass movement, exclusive of rock fall, requires water, and as *PE* decreases the importance of mass movement also decreases. Wind effectiveness varies inversely with *PE*. As precipitation decreases, vegetation becomes sparse, and wind erosion increases. These various combinations are the criteria used to define the morphogenetic regions (Peltier, 1950).

Some of these generalizations may be miscon-

strued. For example, as precipitation decreases, one would think that erosion by water would decrease. However, vegetation also decreases with a decrease in precipitation. With less protection from ground cover, gullying, arroyo cutting, and hillslope erosion may be severe in high-intensity storm rainfall even if annual rainfall is relatively low. This combination can provide similar erosion effects regardless of regional *PE* or *TE* index and is important in "accelerated erosion" (Ruhe and Daniels, 1965).

Past climate

To understand the impact of climate on vegetation, weathering, soil formation, and geomorphic processes requires an evaluation of past climates and how they may have affected a given place. Glacial moraines 13,000 to 14,000 years old in Iowa are direct evidence of a colder environment in the past. The study of past climates is *paleoclimatology*, which may be analyzed theoretically, by botanical or zoological distributions, and by earth-science evidence. With a coordinated attack from all of these fields the environments of a region such as the central Great Plains can be reconstructed (Dort and Jones, 1970).

Theoretical reconstruction

Since glaciation was the most drastic change affecting large areas of the earth during the near past, it is a departure point for paleoclimatic construction. The basis for studying late glacial and postglacial climatic patterns is present-day modal air-mass frontal positions that closely match boundaries of biotic regions in North America (Bryson and Wendland, 1967; Bryson et al., 1969; 1970). Fronts are located as boundaries of air masses that occupy regions more than 50 percent of the time [fig. 11.5(a)]. Maritime tropical air occupies the region south of the modal position of the Pacific winter front. Arctic air occupies the

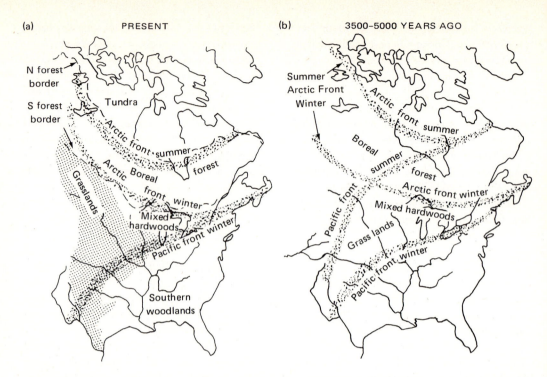

Figure 11.5 (a) The coincidence of biotic regions and meteorologically defined climatic regions. (b) Reconstructed mean frontal and biotic zones 3500 to 5000 years ago. From Bryson and Wendland (1967), by permission of University of Manitoba Press, Winnipeg.

region north of the Arctic summer front, and mild Pacific air occupies the embayment between the Arctic and Pacific winter fronts. Modified Pacific and Arctic air mass is bordered by the Arctic summer and winter fronts. Note the coincidence of tundra and modal Arctic air, and southern woodlands and modal maritime tropical air. Grassland generally conforms to embayed Pacific air east of the Rocky Mountains; mixed hardwoods and boreal forest, to mixed Pacific and Arctic air masses.

The position of the glacier-ice border must be added as a factor affecting climate in late- and postglacial times (Bryson et al., 1969). The border of the ice sheet at any time is based on radiocarbon dating, coastline location, moraines, and other field evidence (fig. 11.6). The summer position of the Arctic front would be along the edge of

the ice sheet, because the strong contrast of the ice in the north with the darker land in the south would fix the main baroclinic zone and the mean front near the ice-land contact. In winter the ice sheet would be a barrier holding very cold Arctic air but permitting southerly outbreaks (Bryson and Wendland, 1967).

Examine the frontal positions that probably existed when ice covered the northern midcontinent 10,000 to 13,000 years ago [fig. 11.7(b)]. The arctic and polar fronts were displaced approximately 10° of latitude farther south than they are at present [fig. 11.5(a)]. The boreal forest and other biotic zones coincidentally shifted with the fronts. Similar reconstructions are possible for 3500 to 5000 years ago [fig. 11.5(b)] and about 8000 years ago [fig. 11.7(a)] relative to the ice front (fig. 11.6). There would be northward displacement of the frontal and biotic zones during the later times. From 13,000 to 20,000 years ago northern air masses and biotic zones probably occupied the upper Mississippi River basin. Knowing the impact of climate on vegetation and soils, we may test these theoretical climatic reconstructions with fossil flora and soil properties—which we will do later.

Air-mass analysis also can be applied locally for paleoclimatic reconstruction. Consider dry-bed pluvial Lake Estancia in central New Mexico and the past climate (Leopold, 1951). There current mean winter and summer temperatures are 30°F and 70°F, with a maximum and minimum of 102°F and −33°F. Average annual rainfall is about 13 inches. The dry bed in the past held a lake 150 feet deep whose surface area was 450 square miles. Negating subsurface leakage or outflow, what climate supported the lake?

Let us approach the problem of past temperatures by considering meridional snow lines and cirques. The present snowline in the Rocky Mountains almost coincides with the mean level of the July freezing isotherm. From the mean lapse rate of temperature (rate of decrease with elevation), the July mean temperature would decrease 11° F for each lowering of the snow line 3280 feet (1000 m). The current mean July freezing isotherm is about 16,500 feet in the atmosphere above Estancia basin. On Sierra Blanca,

Figure 11.6 The borders of the Laurentide ice sheet 20,000 to 3000 years ago, based on radiocarbon dating, coastline locations, and moraines and other field evidence. From Bryson et al. (1969), by permission of Inst. Arctic and Alpine Resources.

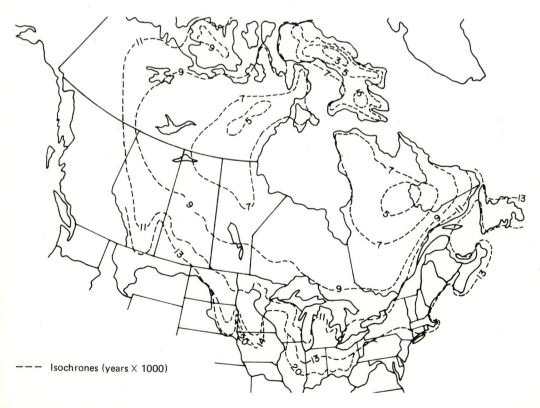

- - - - Isochrones (years X 1000)

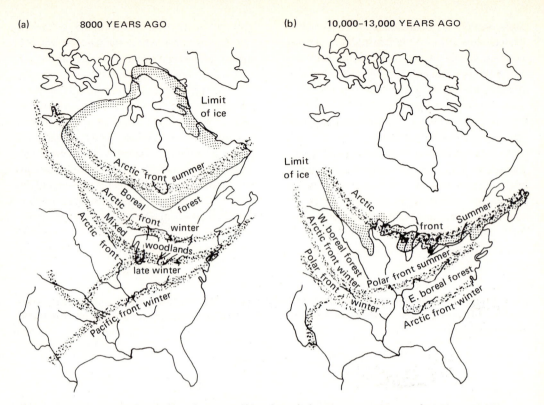

(a) 8000 YEARS AGO

(b) 10,000–13,000 YEARS AGO

Figure 11.7 Reconstructed mean frontal zones and biota boundaries 8000 years ago (a) and 10,000 to 13,000 years ago (b). From Bryson and Wendland (1967), by permission of University of Manitoba Press, Winnipeg.

about 100 miles to the south, glacial cirque floors are about 11,500 feet in elevation.

At Estancia basin let us assume that the snow line was lowered 4500 feet. The mean July temperature would decrease 16° F, the January temperature 5° F, and intervening months proportionately (Leopold, 1951). The mean annual temperature would reduce 12° F. The current annual evaporation from a lake surface in the area is 44 inches, but with a lower annual temperature, the calculated evaporation would be 26 inches.

The hydrologic balance of pluvial Lake Estancia was $e = p + r$ where e is annual evaporation in inches; p, annual rainfall on the lake in inches; and r, runoff from the lake basin catchment in inches. Solving this equation, we find that an an-

nual rainfall of 21 inches is needed to maintain the lake at its highest level. If past temperatures were the same as those today, an annual rainfall of 30 inches would have been required to maintain the balance. This kind of analysis is generally applied to the pluvial lakes of the Southwest (Chapter 9) to evaluate paleoclimatic implications.

Examine a local orographic-convection cell on an oceanic island (Ruhe, 1964b). Assuming that the northeasterly trade winds of the past differed little from those today and that the major topographic form was similar to that of the present, the paleoclimate may be theoretically estimated. The present pattern on Oahu, Hawaii, shows maximum rainfall from northeasterly trade winds just leeward of the Koolau summit line with progressively less rainfall both windward and leeward (fig. 11.8).

To calculate rainfall relative to elevation, fit a network to the summit line of the Koolau Range and windward and leeward down transverse interfluve axes (fig. 11.9). The present rainfall relates exponentially to elevation along the summit line and to distance leeward of the summit line into the Wahiawa basin. The summit (S) relation is:

$$\log R = 1.4812 + 0.00042\ E$$

where R is rainfall in inches and E is elevation in feet. Down the leeward flank into the Wahiawa basin the sector relations (F1 and F2) are:

$$\log R = 2.2743 - 0.0066\ D$$
$$\log R = 2.0753 - 0.0051\ D$$

where D is distance in miles from the summit line.

With higher or lower sea level around Oahu in the past (Chapter 9), the Koolau summit line would have been lower or higher with corresponding less or more summit rainfall respectively. From equation S calculate the rainfall for lower and higher summit lines. From equations F1 and F2 calculate downflank rainfall patterns when the summit line was lower and higher. Plot the calculated values, and construct paleoclimatic maps for the lower and higher sea stands. Lowering sea level increases rainfall, and raising sea level decreases rainfall at a given place. On the Ewa Plain, where present rainfall is 20 inches (fig. 11.8), it could have been 40 inches, or 100 percent more, during low sea level and less than 10 inches, or 50 percent less, during emergence from the high sea stand. Rising up the flank of the Koolau Range, rainfall progressively changes to 25 percent more than present during low sea stand and 25 percent less than present during high sea stand.

These different rainfall systems should have affected erosion, weathering, and soil formation on

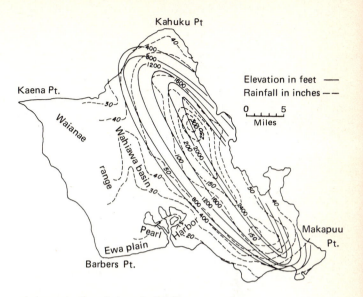

Figure 11.8 The relation of rainfall pattern to the topography of Koolau Range, Oahu, Hawaii. Topographic contours smoothed by computer-calculated least squares fit of ellipses of the general form $aX^2 + bXY + cY^2 + dX + eY + f = 0$.

landscapes that existed during the times of the different sea levels (figs. 7.17, 7.19). They also should have caused different biotic distribution on the island, and they did. A fossil flora is contained in volcanic tuff which was deposited near Pearl Harbor during the period of lower sea level. The flora, collected in an area now receiving less than 20 inches of annual rainfall, has 15 species of plants that characterize the rain belt (Hay and Ijima, 1968). During the low sea stand, rainfall in this area could have been 40 to 70 inches annually! Coincidence of climatic and biotic zones permits testing and proving paleoclimatic distributions by fossil flora.

Paleobotany

Fossil flora is preserved in sediments as macro- or microspecimens, Detrital logs or trees may be rooted in place in such density that they form a

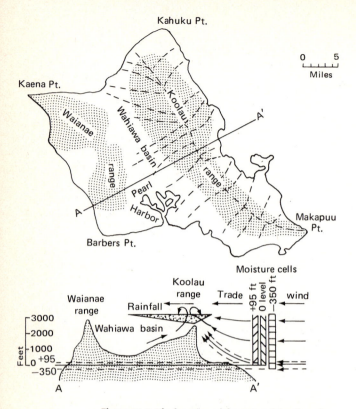

Figure 11.9 The location of the network for the calculation of rainfall relative to the topography of Oahu, Hawaii, and a cross section of the orographic-convection cell. From Ruhe (1964b, 1965b), by permission of Am. J. Sci. © Williams & Wilkins Co., Baltimore.

forest bed. Microflora consists of seeds, spores, and pollen, and the study of it is *palynology*. Wood samples are identified by their structure, which is studied in thin sections under the microscope.

There is a macrofossil record in Iowa that is referred to positions of the glacier fronts in the past (fig. 11.6). From more than 20,000 until 10,000 years ago, Arctic air masses and coincident boreal forest should have dominated (fig. 11.7). Examine the data for wood samples from sediments and the land surface of that time (table

11.1). All the samples from 24,500 to 11,120 years ago by radiocarbon dating are coniferous woods of fir, hemlock, larch, spruce, and yew, which are not native in the state today. The nearest similar forests are now in northern Minnesota, where the environment is cooler and moister (Ruhe, 1969a).

In the more northern area above, present average winter temperatures are 0° to 16°F, and the average summer temperatures are 60° to 66°F. In Iowa these ranges are 14° to 24°F and 72° to 76°F. The northern zone is about 11°F cooler in both winter and summer. The average annual rainfall in the two zones does not differ greatly, being 22 to 32 inches in the north and 26 to 36 inches in the south. But the rainfall distributions differ radically. About 70 percent of the precipitation in Iowa is during the warm months, but prolonged droughts are common during all seasons. In the northern spruce area, precipitation is distributed throughout the growing season. Droughts are occasional but are not so severe, and evaporation is less than it is in Iowa. These contrasting environmental conditions must have existed in Iowa during the past.

Similar wood samples relate to former glacial boundaries during comparable times in Ohio (Goldthwait, 1958; Burns, 1958). There spruce, tamarack, and white cedar occupied the landscape in the past, but today these species exist to the north in Canada.

Pollen analysis is a kind of paleontology in that fossil pollen grains contained in sediments are quantitatively related, in a complex way, to the vegetation of the surrounding region (M. B. Davis, 1963, 1969). Interpretations of fossil pollen are based on relations between modern pollen and modern vegetation. At any one site, percentages of pollen are not equal to the percentages of plants in the vegetation, because of the differences in the amount of pollen produced by different species, the differences in the wind dispersion of different pollen, the differences in the resistance to decay and the reworking of materials

Table 11.1 Woods from Iowa

Sample*	C¹⁴ Date	Wood	County
W-701	250	Box elder	Harrison
W-799	1100 ± 170	Walnut	Harrison
W-699	1800 ± 200	Willow	Harrison
I-2334	1830 ± 100	Red elm	Appanoose
W-702	2020 ± 200	Red elm	Harrison
I-1421	2080 ± 115	American elm	Tama
W-700	11,120 ± 440	Spruce	Harrison
W-882	11,600 ± 200	Spruce	Harrison
I-1019	11,635 ± 400	Spruce	Hamilton
I-1862	11,880 ± 170	Larch	Bremer
C-596	11,952 ± 500	Hemlock	Story
C-912	12,120 ± 530	Hemlock	Webster
C-563	12,200 ± 500	Hemlock	Story
I-2333	12,700 ± 290	Spruce	Linn
W-626	12,970 ± 250	Larch	Hancock
C-913	13,300 ± 900	Hemlock	Webster
W-513	13,820 ± 400	Spruce	Greene
I-1268	13,900 ± 400	Spruce	Hamilton
W-517	13,910 ± 400	Spruce	Greene
C-664	14,042 ± 1000	Hemlock	Story
I-1402	14,200 ± 500	Spruce	Story
W-881	14,300 ± 250	Spruce	Harrison
W-512	14,470 ± 400	Fir, hemlock, larch, spruce	Greene
W-153	14,700 ± 400	Hemlock	Story
I-1270	16,100 ± 1000	Siruce	Boone
I-1024	16,100 ± 500	Spruce	Polk
C-528	16,367 ± 1000	Hemlock	Story
W-126	16,720 ± 500	Spruce, yew	Polk
W-879	19,050 ± 300	Spruce	Harrison
I-1023	21,360 ± 850	Spruce	Pottawattamie
W-141	24,500 ± 1500	Larch	Pottawattamie

From Ruhe (1969a), by permission. © Iowa State University Press, Ames.
*C = University of Chicago; I = Isotopes, Inc.; W = U.S. Geological Survey.

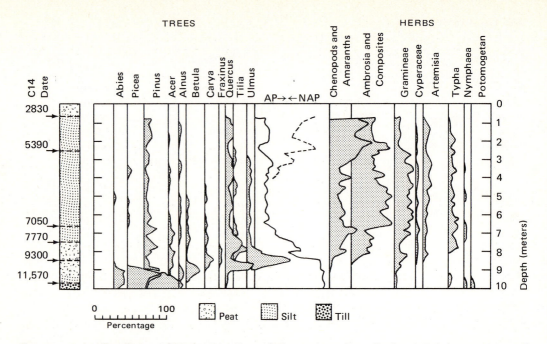

Figure 11.10 A pollen diagram of Woden Bog, Hancock County, Iowa. From Durkee (1971), by permission of Duke University Press, Durham.

containing pollen. Despite the lack of a one-to-one correlation, pollen in sediment is related to plant frequencies, and regional vegetation patterns can thus be delineated.

Pollen analysis consists of microscopically identifying forms and tabulating the numbers and kinds of pollen in samples collected in a vertical section from peat or other sediment. The results are given in a *pollen diagram,* in which the percentage of each pollen species is plotted against depth in a pollen profile (fig. 11.10). In addition, arboreal pollen (AP) and nonarboreal pollen (NAP) are summarized, and the stratigraphic column is diagrammed with radiocarbon dates located if available. In a bog near Woden, Iowa (fig. 11.10), spruce *(Picea)* and fir *(Abies)* dominated the immediate postglacial sediments from 11,500 until about 10,000 years ago (Durkee, 1971). They were followed by birch *(Betula),* alder *(Alnus),* and pine *(Pinus)* from about 10,000 to

9000 years ago. Oak *(Quercus),* basswood *(Tilia),* and elm *(Ulmus)* then were dominant to about 8000 years ago. Note that this last pollen zone coincides with a lower peat buried in the bog. Grass (Gramineae) and herbs (chenopods, amaranths, ambrosia, and composites) became dominant at this time and remained so to the surface of the bog, or to the present. The dual peat over silt stratigraphy is very common in bogs in Iowa, and in all cases arboreal pollen is associated with the lower pair of beds and grass-herb pollen with the upper pair of beds (Walker, 1966; Brush, 1967).

From the base upward, the pollen profile shows coniferous forest followed by deciduous forest and grassland. This sequence is well documented in the midcontinental plains of the United States, where spruce waned about 12,500 years ago in Kansas and Missouri, about 10,000 years ago in Iowa, and 9000 to 9500 years ago in Minnesota (Wells, 1970; Wright, 1970). Prairie began about 9000 years ago in the southwestern part of the area and 7000 to 8000 years ago in the northern area.

Pollen profiles also serve useful purposes locally. In Iowa, hillslopes descend to the double bog sediments (Chapter 6). Peats reflect times of hillslope stability, when organic matter accumulated in the bog faster than mineral sediment was supplied from the bounding hillslopes. Bog silts reflect times of hillslope instability. Arboreal pollen is contained in the buried peat and silt, but a record of trees which would be indicated by weathering and soils is missing on the hillslopes. A layer thick enough to have included any soil profile formed under forest was eroded from the hillslope and formed the upper bog silt (table 6.3). During the last stability episode on the hillslope, soils formed under grassland, as shown by the herb-grass pollen in the upper two layers of the bog. There are no vestiges of forest paleosols on the hillslopes.

Soils and paleosols

As noted previously, paleosols are soils that formed or began forming on a landscape in the past, and there are basically three kinds. *Buried soils* formed and were covered by younger sediment or rock. They are exposed in natural or man-made excavations. *Relict soils* began forming on a pre-existing landscape but were never buried by younger material. Their formation dates from the time of the original landscape and continues today. *Exhumed soils* formed and were buried but were re-exposed on the land surface by erosion of the covering material (Ruhe, 1965a).

Recognition and study

Paleosols are recognized by the same kind, arrangement, and distribution of features as those in soils on the surface (Chapter 2). There is no problem in recognizing a buried soil as a paleosol. It is older than the sediment that buried it. Relict soils require verification that the pre-existing landscape is old and has been stable and subject to little erosion since its formation, and that the soil on it dates from the time of origin of the surface. Exhumed soils are identified by tracing a soil under a sediment and demonstrating that where it is uncovered the mantle was stripped.

Re-examine the polygenetic Red Desert soil from southern New Mexico (fig. 2.13, table 2.7), which occurs on the Picacho surface of the stepped sequence along the Rio Grande (Chapter 7). Organic carbon in the *B* horizon is 9550 years old and was later sealed by carbonate in a *K* horizon. A younger Fillmore surface, less than 2620 years old, is inset below the Picacho surface, which demonstrates that the higher surface is relict and that the soil in question is on the relict surface. The Red Desert soil has all of its horizons with normal clay, iron oxide, and organic carbon distributions to depth. The lower parts of these distributions are engulfed by the carbonate.

Let us consider the paleoclimatic reconstruction of pluvial Lake Estancia, which is 170 miles to the north in central New Mexico. There, where present rainfall is 13 inches, it could have been 21 to 30 inches during the last glacial episode more than 9550 years ago. Thus relict features of the Red Desert soil on the Picacho surface, namely clay, iron oxide, and organic carbon profiles, can relate to a past wetter environment. Engulfment of the lower part of the profile by carbonate after 9550 years can relate to more arid conditions of postglacial time.

Note that morphological, physical, and chemical properties are used to study this relict soil. In some situations the chemistry of buried and exhumed soils is meaningless. Where buried, some paleosols may be secondarily enriched by bases and metals brought down during weathering of the overlying sediment. Some buried soils under loess in the Midwest have pH values of over 6.0 and high base saturations, whereas analogous ground soils may have pH values of 4.5 to 5.5 and base saturations of 40 percent in comparable horizons.

Where chemical analysis is not feasible, studies of physical and mineralogic properties are useful. Texture and structure will identify the stage of soil development. Weathering ratios of grain and clay-size minerals help quantify the paleosol system and are useful in environmental interpretation.

Environmental implications

If paleosols are studied in comparison with ground soils, whose environments are better understood, certain interpretations may be made about the environments in which the paleosols formed (Ruhe, 1970). Compare the gross morphologies of a paleosol with ground soils in Iowa (fig. 11.11). Profile A is under grass, and profile C is under forest (table 2.6). Paleosol profile B is

Figure 11.11 Soils and paleosols in Iowa. (a) Tama soil formed under grass. (b) Late Sangamon paleosol. (c) Fayette soil formed under trees. Compare morphology of (b) with that of (a) and (c). Photographs (a) and (c) by R. W. Simonson, by permission. From Ruhe (1970), by permission of University Press of Kansas, Lawrence.

a   b   c

Figure 11.12 A road cut near Kapaa, Kauai, Hawaii: A1 and A2, younger and older alluvium; B, basalt; P1 and P2, younger and older buried soils. V, volcanic ash.

more like soil profile *C* than *A,* so the paleosol formed under trees rather than grass.

This kind of paleosol is buried beneath loess from Illinois westward through southern Iowa and at least 70 miles west of the Missouri River in Kansas and Nebraska. Throughout this prairie region, ground soils formed under grass in the moist to dry climatic zones (fig. 11.1). The forest paleosol shows that environmental conditions were drastically different from those today and required a humid climate.

Currently the nearest northern forest is 350 miles northeast, the nearest southern forest 175 miles southeast, of the Kansas-Nebraska plains area, so the temperature regime is difficult to evaluate. Other paleosols on the paleogeomorphic surface change from forest to grassland soils, Chernozems, and semiarid soils from northeast to northwest Kansas. The paleosol great-group boundaries are offset about 100 miles to the west in contrast to those of the ground soils (Frye and Leonard, 1957).

Not all buried soils differ in kind from the ground soils that overlie them. Examine a complex roadcut near Kapaa, Kauai, Hawaii (fig.

11.12). From the base upward, a soil *P2* formed on volcanic ash *(V)* and was buried by basalt lava *(B).* The basalt was eroded and covered by alluvium *A2* without intervening weathering or soil formation. A second soil, *P1,* formed on alluvium *A2* and was later buried under alluvium *A1* on a hillslope. Buried soil *P2* has color (red), texture, and structure similar to those of the ground soils in the area, and its composition is essentially sesquioxide and clay. At a site beneath the basalt *(B),* the paleosol *B* horizon has 16 to 19 percent free iron oxide. By x-ray diffraction analysis with MoKα radiation, the mineral composition is seen to be dominantly kaolinite and hematite with gibbsite, goethite, and maghemite. This buried soil is similar in kind to the ground soils of the area (table 2.8), so both soils should have formed under a similar environment. But, as pointed out previously, the environment in Hawaii changed with the rise and fall in sea level. The problem is complex!

Soil structures may also be used in reconstructing past environments. When glacier-ice sheets occupied specific parts of the midcontinental United States at specific times in the past, permafrost should have affected the land just beyond the ice, and patterned ground should have formed. Many fossil features of patterned ground (fig. 11.3) occur in this region (R. F. Black, 1969; Frye and Willman, 1958; Ruhe 1969a).

Climatic estimates for the midcontinent region may be made by comparison with the present cold regions elsewhere. In Alaska today active and inactive ice wedges correlate with climatic zones that are aligned with latitude (Péwé, 1966). If similar conditions prevailed in the midcontinental United States during glacial times, a $-2.5°C$ isotherm should have crossed the region now known as central Illinois, Indiana, and Ohio about 20,000 years ago (Wayne, 1967). A $+5°C$ isotherm should have crossed the region where the Ohio and Missouri River Valleys are today. As the ice margin retreated northward from 20,000 to 10,000 years ago, the peripheral permafrost zone should also have been displaced northward to the region now occupied by northern Minnesota, Wisconsin, and Michigan (fig. 11.6).

An array of pollen profiles from the region that have been radiocarbon dated shows that climatic patterns like those of the present developed about 8000 years ago. These changing conditions should have affected landscapes and soils during late and postglacial times, and any geomorphic or pedologic studies of the region must consider the possibility of these environmental changes.

Past environments may also be reconstructed from invertebrate and vertebrate faunas contained in surficial deposits, but these fields of study are best handled by their specialists. In all of these approaches that evaluate changes from the past to the present, time must be included in the analysis.

Dating landscapes and Surficial deposits

Land surfaces and surficial sediments can be dated by two basic methods. In a *relative* method one feature may be determined as younger, older,

or the same age as another feature. This is only a qualitative approach, as there is no way to determine how great or how small the difference in age may be. In an *absolute* method a natural chronometer is built into the system, and time is measured by radiometric techniques. Hence, one feature will have a specific age, and another feature will have another specific age. The difference in age can be quantified as number of years. In most studies, there are not enough absolute dates available, so a combination of relative and absolute methods must be used to establish a chronology for a system. This kind of study is termed *geochronology*.

Relative methods

If surficial deposits are horizontal or deviate slightly from the horizontal, the *law of superposition* applies, and younger beds are on older beds. Note the upward decrease in radiocarbon ages of beds in the Woden Bog (fig. 11.10). Distortion or deformation of beds causes complex arrangements which must be deciphered structurally to determine the relative ages of the beds.

The land surface must be fitted to surficial deposits. Consider a stream valley incised in horizontal beds. Younger alluvium in the stream valley is below the elevation of an older bed just beneath the land surface under the adjacent hilltop. Younger beds may be inset within the alluvium in a terrace relationship (Chapter 4). In this valley-incised landscape, the hillslope deviates from horizontal and descends across beds or zones within the hill, so the hillslope is younger than the uppermost bed in the hill. The plane of the hillslope angularly intersects the plane of the hill summit, which shows that the hillslope is younger. The hillslope descends to the alluvial fill in the valley and must be the same age as the alluvium because erosion on the hillslope provided the sediment at the footslope.

The age of one land surface relative to another or to a body of surficial sediment is readily determinable if the land surface is recognized by its true geometric nature in space. It is a plane which can vary from a simple two-dimensional linear form to a complex curvilinear surface (fig. 6.1). Where the hillslope descends to the alluvial fill, a plane representing the hillslope surface splits into two planes at the edge of the alluvium. One plane passes under the alluvium; the other passes across the top of the alluvium. The maximum and minimum ages of the hillslope are represented by the base and top of the alluvium respectively.

The principles of relative dating evolve from the hillslope relations. A land surface is younger than the youngest deposit that it cuts. A land surface is younger than the youngest land surface that it cuts. A land surface is contemporaneous with alluvial deposits that lie on it. *Principles of ascendancy and descendancy* apply to the hillslope proper. It is younger than the higher surface to which it ascends but is the same age as the sediments to which it descends.

Where erosion surfaces cut cross country, several self-evident relations determine their ages (Trowbridge, 1921). Any erosion surface (1) is younger than the youngest material that it cuts; (2) is younger than any structure that it bevels; (3) is younger than any material of which there are distinguishable fragments or fossils in alluvial deposits on the surface; (4) is contemporaneous with alluvial deposits which lie on it; (5) is the same age as or is older than other terrestrial deposits lying on it; (6) is older than valleys cut below it; (7) is younger than erosion remnants above it; (8) is older than deposits in valleys below it; (9) is younger than any erosion surface standing at a higher level; and (10) is older than any erosion surface standing at a lower level.

Absolute methods

The most widely used methods for absolute dating are based on the radioactive decay of natural-occurring elements, which follows a rate law like a first-order chemical reaction. Exponen-

tial decay is expressed as radioactive half-life, meaning the time required for an element to lose half of its radioactivity. The most common isotope of radium has a half-life of 1620 years. Beginning at time zero, radioactivity would be half the original value at 1620 years, a quarter of the value at 3240 years, an eighth at 4860 years, and so on.

Some of the well-known methods (Kulp, 1961) are uranium-lead (U^{238}-Pb^{206}, U^{235}-Pb^{207}), thorium-lead (Th^{232}-Pb^{208}), rubidium-strontium (Rb^{87}-Sr^{87}), and potassium-argon (K^{40}-Ar^{40}). These methods are used to determine the ages of rocks containing minerals such as uraninite (pitchblende), zircon, micas, and potash feldspar. A Th^{230}-U^{234} method is used to date fossil coral (Thurber et al., 1965; Broecker and Thurber, 1965); a Pa^{231}-Th^{230} method, to date sediments of deep-sea cores (Rosholt et al., 1961).

These methods do not apply directly to terrestrial surficial sediments, which may contain detrital minerals and elements of any age, but can date rocks to which land surfaces or surficial deposits are related. On Oahu, Hawaii, the Koolau volcanic series is dated by the K^{40}-Ar^{40} method as 2.6 to 1.8 million years old (Doell and G. B. Dalrymple, 1973). Eight erosion surfaces of the stepped sequence from Mahoe to Kamana (Chapter 7) are cut in the Koolau basalts and are younger. In the Sierra Nevada in California a volcanic tuff dated by the K^{40}-Ar^{40} method as about 700,000 years old rests on Sherwin glacial till, which must be older. A basalt flow dated as 60,000 to 90,000 years old underlies Tahoe glacial till, which in turn must be younger (G. B. Dalrymple, 1964). By relative association a specific minimum or maximum age can be applied to the surficial deposits, but they can be somewhat or much younger or older than the associated radioisotope age.

At present only the radiocarbon method (Libby, 1965) directly applies to the absolute dating of surficial deposits, and it reaches only 40,000 years into the past. Radioactive C^{14} with a half-life of 5750 years is produced by cosmic rays

Table 11.2 Effects of varying amounts of modern and old carbon on true radiocarbon age

True age (yr)	Apparent age after contamination by old carbon* (yr)			
	5%	10%	20%	50%
500	900	1300	2200	6000
5000	5400	5800	6700	10,500
10,000	10,400	10,800	11,700	15,500
20,000	20,400	20,800	21,700	25,500

True age (yr)	Apparent age after contamination by modern carbon (yr)			
	1%	5%	20%	50%
5000	4950	4650	3700	2100
10,000	9800	9000	6800	3600
20,000	19,100	16,500	10,600	5000
30,000	27,200	21,000	12,200	5400

From Polach and Golson (1966), by permission of Aust. Inst. Aboriginal Stud.
*Old carbon where radioactivity is almost gone.

bombarding N^{14} in the atmosphere: $N^{14} + n = C^{14} + H^1$. The C^{14} combines with oxygen, forming radioactive CO_2, which mixes with atmospheric CO_2. Plants take up CO_2 in photosynthesis and become radioactive. Animals eat plants and also become radioactive.

During life all organic matter maintains an equilibrium, with all radioactive decay being replaced by C^{14} in photosynthesis or food supply. Specific activity is kept at a level of about 14 disintegrations/min/g: $C^{14} = \beta^- + N^{14}$. The C^{14} atoms are assimilated by organisms at the rate that C^{14} atoms disintegrate to form N^{14}. Upon death, assimilation stops, and decay begins, with half of the specific activity lost after 5750 years, another half after another 5750 years, and so on. Measurement of the specific activity of a sample yields its radiocarbon age.

Radiocarbon age and true age differ (Stuiver and Suess, 1966), and conversion involves assumptions about the half-life of C^{14}, the production rate of C^{14} by cosmic rays, the size of the C^{14}

reservoir, and the exchange rate of the distribution. In the original work on C^{14} by Libby in 1947, these relations were assumed to be constant, but they are now known to vary. Discrepancies occur between radiocarbon and calendar ages of wood, and 2000-year radiocarbon ages are generally 50 to 100 years greater than true ages. For the calendar year 1800, radiocarbon ages are generally 20 years less than the true age. The discrepancy between true and radiocarbon ages increases rapidly with increasing age, but an approximation of true age T can be estimated from radiocarbon age R by the equation $T = 1.4 R - 1100$.

The natural contamination of an organic sample is another problem in radiocarbon dating. Many samples are buried in an environment of foreign carbonaceous material in the form of humus, plant rootlets, or organic solutes such as humic and fulvic acids. Younger solutes from a ground soil can descend into the Ab horizon of a buried soil. Older material can contaminate younger material, such as black-shale fragments in a peaty muck. The more discrepant the ages of true sample and contaminant are, the greater is the error in the age of the sample (table 11.2).

The radiocarbon dating of soil organic carbon and the interpretation of results demand special care. The components of soil humus, defined on the basis of simple chemical extraction, are humin, humic acid, and fulvic acid (fig. 2.3). Prior to dating, the standard pretreatment of a sample in the laboratory consists of hand removal of obvious foreign matter such as rootlets, and the removal of inorganic carbonate contaminants using hydrochloric acid. Consider a soil A horizon subjected to various chemical treatments (Campbell et al., 1967 a, b) and the resulting radiocarbon dates (fig. 11.13). The original raw sample is 870 ± 50 years old, but the hydrochloric acid extract is 325 ± 60 years old. During the standard laboratory pretreatment to remove carbonates, then, an alteration in the radiocarbon system begins.

Where contamination of a sample is also sus-

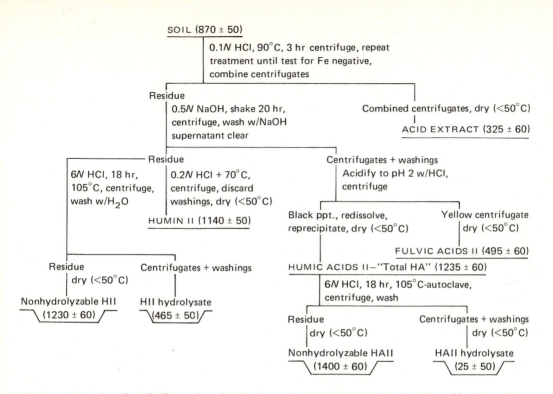

Figure 11.13 Fractionation of soil organic carbon by chemical treatment and radiocarbon dates of fractions. From Campbell et al. (1967a, b), by permission. © Williams & Wilkins Co., Baltimore.

pected, a second pretreatment uses sodium hydroxide to remove adsorbed humic acids (fig. 11.13). The sample is fractionated into residue, or humin, and NaOH-soluble humic and fulvic acids. These dated fractions in order are 1140 ± 50, 1235 ± 60, and 495 ± 40 years old. With other extracts, the dates range from 25 ± 50 to 1400 ± 60 years. All of these dates are from a raw sample whose age was 870 years, and these variations must be kept in mind when one interprets the radiocarbon dates of soil organic carbon. In general, if the humin and humic acid dates reasonably agree, contamination is not significant, and the dates are considered valid.

The variability within a buried-soil A horizon may be great. From a paleosol in Australia, hand-separated carbonized specks were 33,700 $^{+\,2200}_{-\,1730}$ years old. A fine-earth fraction containing organic carbon was 19,980 ± 370 years old, with its NaOH-soluble fraction being 24,960 ± 580 years old and its NaOH-insoluble fraction 25,360 ± 580 years old. The sample is contaminated (Polach et al., 1969). Generally humus is concentrated in the clay-size fraction of soil and is separable by the sedimentation technique (Scharpenseel et al., 1968). For comparison purposes, all radiocarbon dating of soil organic carbon should be standardized to the fine-earth fraction.

Another study of soil organic carbon extracts is useful for environmental interpretation. Infrared absorption spectra of humic acids distinguish between soils formed under grass and those formed under trees. The spectral diagrams relate to the molecular weight of the substance examined and to the vegetation under which the humic acids formed or were transformed (Dormaar, 1967). This technique is applied in paleopedology (Dormaar and Lutwick, 1969).

Radiocarbon dating is also applied to inorganic carbon in carbonate, but great care must be used in interpreting the dates. Recall solution and precipitation in the carbonate–carbon dioxide–water system (Chapter 2):

$$H_2O + CO_2 + CaCO_3 \rightleftharpoons 2HCO_3^- + Ca^{++}$$

During the solution phase, young carbon from CO_2 in soil water combines with old carbon in the pre-existing carbonate to form bicarbonate. Upon reversal of the process, there is no selective precipitation of old and young carbon in the secondary carbonate, and about a half is old and a half is young carbon. Radiocarbon dating measures the mix.

Other techniques are also used for absolute dating. *Dendrochronology* is the systematic counting of tree rings to establish the age of forest stands and the minimum age of soils, landslides, terraces, and glacial moraines (Fritts, 1965). Age estimates are accurate only if a single ring is formed each year. Climatic changes may be interpreted from variations in ring widths. *Lichenometry* is a method for dating exposed rock surfaces or active geologic processes in treeless areas by measuring the rate of lichen growth (Reger and Péwé, 1969). It is based on the slow, constant increase of individual plant diameters, and after the growth rate is known, age is determined by measuring the diameters of lichen thalli growing on a critical surface.

If materials for absolute dating are in surficial sediments or are associated with land surfaces, time benchmarks can be established. By employing the principles of relative dating to the benchmarks, a chronology can be developed for the entire system of geomorphic processes, landscapes, surficial deposits, and soils. Not only can the history of the earth be thus placed in perspective, but estimates can be made of future change. In terms of how these sciences are put to practical use, these are what geomorphology and surficial geology are all about!

Bibliography

Alden, W. C., and M. M. Leighton (1917). The Iowan drift: a review of the evidences of the Iowan stage of glaciation. *Iowa Geol. Surv. Annu. Rept.* 26:49-212.

Allen, B. L., and E. P. Whiteside (1954). The characteristics of some soils on tills of Cary and Mankato age in Michigan. *Soil Sci. Soc. Am. Proc.* 18:203-206.

Allen, R. H. (1972). *A Glossary of Coastal Engineering Terms.* U.S. Army, Corps of Engineers, Coastal Engineering Research Center misc. paper 2-72.

American Commission on Stratigraphic Nomenclature (1961). Code of stratigraphic nomenclature. *Am. Assoc. Petrol. Geol. Bull.* 45:645-665.

Arkley, R. J., and H. C. Brown (1954). The origin of mima mound (hogwallow) microrelief in the far western states. *Soil Sci. Soc. Am. Proc.* 18:195-199.

Armstrong, L. C. (1940). Decomposition and alteration of feldspars and spodumene by water. *Am. Mineral.* 25:810-820.

Bagnold, R. A. (1941). *The Physics of Blown Sand and Desert Dunes.* Methuen & Co., London.

Bakker, J. P., and J. W. N. LeHeux (1946). Projective geometric treatment of O. Lehmann's theory of transformation of steep mountain slopes. *Proc. Kon. Ned. Akad. Wetensch.* 49:533-547.

_____ (1947). Theory on central rectilinear recession of slopes I & II. *Proc. Kon. Ned. Akad. Wetensch.* 50:959-966, 1154-1162.

_____ (1950). Theory on central rectilinear recession of slopes III & IV. *Proc. Kon. Ned. Akad. Wetensch.* 53:1073-1084, 1364-1374.

_____ (1952). A remarkable new geomorphological law I, II, & III. *Proc. Kon. Ned. Akad. Wetensch.* 55:399-410, 554-571.

Bakker, J. P., and A. N. Strahler (1956). *Report on Quantitative Treatment of Slope Recession Problems.* Int. Geogr. Union 1st Rept. Comm. Slopes, Rio de Janeiro:30-41.

Barnes, C. P. (1948). Environment of natural grassland. In *Grass, U.S. Department of Agriculture Yearbook of Agriculture,* pp. 45-49.

Barnett, A. P., and B. H. Hendrickson (1960). Erosion on Piedmont soils. *Soil Conserv.* (U.S. Department of Agriculture) 26:31-34.

Beaty, C. B. (1959). Slope retreat by gullying. *Geol. Soc. Am. Bull.* 70:1479-1482.

_____ (1963). Origin of alluvial fans, White Mountains, California, and Nevada. *Ann. Assoc. Am. Geogr.* 53:516-535.

_____ (1970). Age and estimated rate of accumulation of an alluvial fan, White Mountains, California, U.S.A. *Am. J. Sci.* 268:50-77.

Beavers, A. H., J. B. Fehrenbacher, P. R. Johnson, and R. L. Jones (1963). CaO-ZrO$_2$ molar ratios as an index of weathering. *Soil Sci. Soc. Am. Proc.* 27: 408-412.

Benson, M. A., and T. Dalrymple (1967). General field and office procedures for indirect discharge measurements. *U.S. Geol. Surv. Tech. Water Res. Invest.*, bk. 3, chap. A1.

Benton, T. E., D. S. Gray, F. R. Lesh, and J. E. McKeehen (1925). *Soil Survey of Kossuth County, Iowa,* ser. 1925, no. 19. U.S. Department of Agriculture, Bureau of Chemistry and Soils.

Berg, L. S. (1964). *Loess as a Product of Weathering and Soil Formation.* U.S. Department of Agriculture–National Science Foundation (Translated from Russian).

Berry, L., and B. P. Ruxton (1959). Notes on weathering zones and soils on granitic rocks in two tropical regions. *J. Soil Sci.* 10:54-63.

Birot, P. (1968). *The Cycle of Erosion in Different Climates.* University of California Press, Berkeley.

Black, C. A., ed. (1965a). *Methods of Soil Analysis: Pt. 1., Physical and Mineralogical Properties Including Statistics of Measurement and Sampling.* American Society of Agronomy monograph 9. Madison, Wis.

_____ (1965b). *Methods of Soil Analysis: Pt. 2., Chemical and Microbiological Properties.* American Society of Agronomy monograph 9. Madison, Wis.

Black, R. F. (1969). Climatically significant fossil periglacial phenomena in north central United States. *Biuletyn Peryglacjalny* 20:225-238.

Bluck, B. J. (1964). Sedimentation of an alluvial fan in southern Nevada. *J. Sedimen. Petr.* 34:395-400.

Blumenstock, D. I., and C. W. Thornthwaite (1941). Climate and the world pattern. In *Climate and Man, Yearbook of Agriculture,* pp. 98-127. U.S. Department of Agriculture.

Bradley, W. C. (1957). Origin of marine-terrace deposits in the Santa Cruz area, California. *Geol. Soc. Am. Bull.* 68:421-444.

Bray, R. H. (1937). Chemical and physical changes in soil colloids with advancing development in Illinois soils. *Soil Sci.* 43:1-14.

Brewer, R. (1964). *Fabric and Mineral Analysis of Soils.* John Wiley & Sons, New York.

Broecker, W. S., and D. L. Thurber (1965). Uranium-series dating of corals and oolites from Bahaman and Florida Key limestones. *Science* 149:58-60.

Brophy, J. A. (1959). *Heavy Mineral Ratios of Sangamon Weathering Profiles in Illinois.* Ill. Geological Survey circ. 273.

Brown, R. L., and A. L. Hafenrichter (1962). *Stabilizing Sand Dunes on the Pacific Coast with Woody Plant.* U.S. Department of Agriculture, Soil Conservation Service misc. publ. 892.

Brush, G. S. (1967). Pollen analysis of late glacial and postglacial sediments in Iowa. In *Quaternary Paleoecology,* ed. E. J. Cushing and H. E. Wright, pp. 99-115. Yale University Press, New Haven.

Bryan, K. (1936). The formation of pediments. *Sixteenth International Geological Congress Report,* vol. 2, pp. 765-775. Washington, D.C.

_____ (1940). Gully gravure, a method of slope retreat. *J. Geomorphol.* 3:89-107.

Bryson, R. A., and W. M. Wendland (1967). Tentative climatic patterns for some late glacial and post glacial episodes in central North America. In *Life, Land, and Water: Proceedings of the 1966 Conf. on Environmental Studies of Glacial Lake Agassiz,* ed. W. J. Mayer-Oakes, pp. 271-298. University of Manitoba Press, Winnipeg.

Bryson, R. A., W. M. Wendland, J. D. Ives, and J. T. Andrews (1969). Radiocarbon isochrones on the disintegration of the Laurentide ice sheet. *Arctic and Alpine Res.* 1:1-14.

Bryson, R. A., D. A. Baerreis, and W. M. Wendland (1970). The character of late glacial and postglacial climatic changes. In *Pleistocene and Recent Environments of the Central Great Plains,* eds. W. Dort and J. K. Jones, pp. 53-74. University Press of Kansas, Lawrence.

Buchanan, T. J., and W. P. Somers (1968). Stage measurement at gaging stations. *U.S. Geol. Surv. Tech. Water Res. Invest.,* bk. 3, chap. A7.

_____ (1969). Discharge measurements at gaging stations. *U.S. Geol. Surv. Tech. Water Res. Invest.,* bk. 3, chap. A8.

Bull, W. B. (1963). Alluvial-fan deposits in western Fresno County, California. *J. Geol.* 71:243-251.

_____ (1964). *Geomorphology of Segmented Alluvial Fans in Western Fresno County, California,* pp. 89-129. U.S. Geological Survey prof. paper 352-E.

Burns, G. W. (1958). Wisconsin age forests in western Ohio: II, vegetational and burial conditions. *Ohio J. Sci.* 58:220-230.

Butler, B. E. (1956). Parna: an eolian clay. *Australian J. Sci.* 18:145-151.

Caldwell, R. E., and J. L. White (1956). A study of the origin and distribution of loess in southern Indiana.

Soil Sci. Soc. Am. Proc. 20:258-263.

Campbell, C. A., E. A. Paul, D. A. Rennie, and K. J. McCallum (1967a). Factors affecting the accuracy of the carbon-dating method in soil humus studies. *Soil Sci.* 104:81-85.

_____(1967b). Applicability of the carbon-dating method of analysis to soil humus studies. *Soil Sci.* 104:217-224.

Carlston, C. W. (1965). The relation of free meander geometry to stream discharge and its geomorphic implications. *Am. J. Sci.* 263:864-885.

_____(1969). Downstream variations in the hydraulic geometry of streams: special emphasis on mean velocity. *Am. J. Sci.* 267:499-509.

Carlston, C. W., and W. B. Langbein (1960). *Rapid Approximation of Drainage Density: Line Intersection Method.* U.S. Geological Survey, Water Resources Division bull. 11.

Carson, M. A. (1971). *The Mechanics of Erosion.* Pion, London.

Carson, M. A., and M. J. Kirkby (1972). *Hillslope Form and Process.* Cambridge University Press, London.

Carter, R. W., and J. Davidian (1968). General procedure for gaging streams. *U.S. Geol. Surv. Tech. Water Res. Invest.,* bk. 3, chap. A6.

Chapman, R. W. (1946). Lithification of Pleistocene clay at Kahuku Point, Oahu. *Geol. Soc. Am. Bull.* 57:985-996.

Chen, C. N. (1970). *Removal of a Spherical Particle from a Flat Bed.* Georgia Institute of Technology Environmental Research Center, ERC-0770.

Chepil, W. S. (1959). Equilibrium of soil grains at the threshold of movement by wind. *Soil Sci. Soc. Am. Proc.* 23:422-428.

Chorley, R. J. (1962). *Geomorphology and General Systems Theory.* U.S. Geological Survey prof. paper 500-B.

Chorley, R. J., ed. (1972). *Spatial Analysis in Geomorphology.* Harper & Row, New York.

Chorley, R. J., A. J. Dunn, and R. P. Beckinsale (1964). *The History of the Study of Landforms or the Development of Geomorphology.* Methuen & Co., London.

Ciaccio, L. L., ed. (1973). *Water and Water Pollution Handbook.* Vols. 1-4. Marcel Dekker, New York.

Cline, M. G. (1949). Basic principles of soil classification. *Soil Sci.* 67:81-91.

_____(1955). *Soil Survey: Territory of Hawaii,* ser. 1939, no. 25. U.S. Department of Agriculture, Soil Conservation Service.

Coates, D. R. (1958). *Quantitative Geomorphology of Small Drainage Basins of Southern Indiana.* Columbia University, Department of Geology,

ONR tech. rept. 10, proj. NR 389-042.

Collinet, J. (1969). Contribution à l'étude des "Stone-lines" dans la région du Moyen-Ogooué (Gabon). *Cah. O.R.S.T.O.M. (Paris),* sér. pédol. 7:3-42.

Conover, L. F. (1960). *Documentation of the Storm Surge as Hurricane Donna Crossed the Florida Keys, September 10, 1960.* National Hurricane Research Project, Miami.

Cooper, W. S. (1958). Coastal sand dunes of Oregon and Washington. *Geol. Soc. Am. Memoir 72.*

Corps of Engineers (1968). *Flooding along Clifty Creek in Vicinity of Columbus, Indiana.* U.S. Army.

Cotton, C. A. (1941). *Landscape as Developed by the Processes of Normal Erosion.* Cambridge University Press, London.

_____ (1942). *Climatic Accidents in Landscape Making.* John Wiley & Sons, New York.

Cowie, J. D. (1964). Loess in the Manawatu District, New Zealand. *N. Z. J. Geol. and Geophys.* 7: 389-396.

Crickmay, C. H. (1933). The later stages of the cycle of erosion. *Geol. Mag.* 70:337-347.

Culling, W. E. H. (1963). Soil creep and the development of hillside slopes. *J. Geol.* 71:127-161.

Dalrymple, G. B. (1964). Potassium-argon dates of three Pleistocene interglacial basalt flows from the Sierra Nevada, California. *Geol. Soc. Am. Bull.* 75:753-758.

Dalrymple, T., and M. A. Benson (1967). Measurement of peak discharge by the slope-area method. *U.S. Geol. Surv. Tech. Water Res. Invest.,* bk. 3, chap. A2.

Daniels, R. B. (1960). Entrenchment of the Willow Drainage Ditch, Harrison County, Iowa. *Am. J. Sci.* 258:161-176.

Daniels, R. B., R. L. Handy, and G. H. Simsonson (1960). Dark-colored bands in the thick loess of western Iowa. *J. Geol.* 68:450-458.

Daniels, R. B., M. Rubin, and G. H. Simonson (1963). Alluvial chronology of the Thompson Creek watershed, Harrison County, Iowa. *Am. J. Sci.* 261:473-487.

Daniels, R. B., and R. H. Jordan (1966). *Physiographic History and the Soils, Entrenched Stream Systems, and Gullies, Harrison County, Iowa.* U.S. Department of Agriculture tech. bull. 1348.

Davis, M. B. (1963). On the theory of pollen analysis. *Am. J. Sci.* 261:897-912.

_____(1969). Palynology and environmental history during the Quaternary period. *Am. Sci.* 57: 317-332.

Davis, R. E., and F. S. Foote (1940). *Surveying: Theory*

and Practice. 3d ed. McGraw-Hill, New York.

Davis, W. M. (1899a). The geographical cycle. Geogr. J. 14:481-504.

_____ (1899b). The peneplain. Am. Geol. 23:207-239.

_____ (1954). Geographical Essays. Unabridged republ. 1909 ed. Dover, New York.

Denny, C. S. (1965). Alluvial Fans in the Death Valley Region, California and Nevada. U.S. Geological Survey prof. paper 466.

_____ (1967). Fans and pediments. Am. J. Sci. 265: 81-105.

Denny, C. S., and J. C. Goodlett (1968). Tree-Throw Origin of Patterned Ground on Beaches of the Ancient Champlain Sea near Plattsburgh, New York, pp. 157-164. U.S. Geological Survey prof. paper 600-B.

Dixey, F. (1948). Geology of Northern Kenya. Kenya Geological Survey rept. 15.

Doell, R. R., and G. B. Dalrymple (1973). Potassium-argon ages and paleomagnetism of the Waianae and Koolau Volcanic Series, Oahu, Hawaii. Geol. Soc. Am. Bull. 84:1217-1242.

Donn, W. L., W. R. Farrand, and M. Ewing (1962). Pleistocene ice volumes and sea-level lowering. J. Geol. 70:206-214.

Dormaar, J. F. (1967). Infrared spectra of humic acids from soils formed under grass or trees. Geoderma 1:37-45.

Dormaar, J. F., and L. E. Lutwick (1969). Infrared spectra of humic acids and opal phytoliths as indicators of paleosols. Can. J. Soil Sci. 49:29-37.

Dort, W., and J. K. Jones, eds. (1970). Pleistocene and Recent Environments of the Central Great Plains. University of Kansas Press, Lawrence.

Dreimanis, A. (1959). Measurements of depth of carbonate leaching in service of Pleistocene stratigraphy. Geol. Fören Förh. 81:478-484.

Dryden, L., and C. Dryden (1946). Comparative rates of weathering of some common heavy minerals. J. Sedimen. Petr. 16:91-96.

Dudal, R. (1968). Definitions of Soil Units for the Soil Map of the World. UNESCO, FAO, World Soil Resources rept. 33.

Durkee, L. H. (1971). A pollen profile from Woden Bog in north-central Iowa. Ecology 52:837-844..

Easterbrook, D. J. (1969). Principles of Geomorphology. McGraw-Hill, New York.

Elson, J. A. (1967). Geology of Glacial Lake Agassiz. In Life, Land, and Water, ed. W. J. Mayer-Oakes, pp. 37-95. University of Manitoba Press, Winnipeg.

Embleton, C., and C. A. M. King (1968). Glacial and Periglacial Geomorphology. St. Martin's Press, New York.

Emery, K. O. (1958). Shallow submerged marine terraces of southern California. Geol. Soc. Am. Bull. 69:39-60.

Engineering Experiment Station Staff (1966). Perennial and emphemeral streams and lakes in Indiana. Purdue University map.

Everitt, B. L. (1968). Use of the cottonwood in an investigation of the recent history of a floodplain. Am. J. Sci. 266:417-439.

Fairbridge, R. W., ed. (1968). The Encyclopedia of Geomorphology. Reinhold Publishing Corp., New York.

Fehrenbacher, J. B. (1973). Loess stratigraphy, distribution, and time of deposition in Illinois. Soil Sci. 115:176-182.

Fehrenbacher, J. B., J. L. White, H. P. Ulrich, and R. T. Odell (1965). Loess distribution in southeastern Illinois and southwestern Indiana. Soil Sci. Soc. Am. Proc. 29:566-572.

Fellenius, W. (1936). Calculations of the stability of earth dams. Trans. 2d. Cong. on Large Dams 4, Washington, D.C.

Finkel, H. J. (1959). The barchans of southern Peru. J. Geol. 67:614-647.

Finney, H. R., N. Holowaychuk, and M. R. Heddleson (1962). The influence of microclimate on the morphology of certain soils of the Allegheny Plateau in Ohio. Soil Sci. Soc. Am. Proc. 26:287-292.

Fisk, H. N. (1944). Geological Investigation of the Alluvial Valley of the Lower Mississippi River. U.S. Army, Corps of Engineers, Mississippi River Commission, Vicksburg, Miss.

_____ (1951). Loess and Quaternary geology of the lower Mississippi Valley. J. Geol. 59:333-356.

Flint, R. F. (1947). Glacial Geology and the Pleistocene Epoch. John Wiley & Sons, New York.

_____ (1949). Leaching of carbonates in glacial drift and loess as a basis for age correlation. J. Geol. 57:297-303.

_____ (1957). Glacial and Pleistocene Geology. John Wiley & Sons, New York.

_____ (1971). Glacial and Quaternary Geology. John Wiley & Sons, New York.

Flint, R. F., and Contributors (1945). Glacial map of North America. Geological Society of America spec. paper 60.

_____ (1959). Glacial map of the United States east of the Rocky Mountains, 2 sheets. Geological Society of America.

Fölster, H. (1969). Slope development in SW-Nigeria

during late Pleistocene and Holocene. *Univ. Göttinger Bodenk. Ber.* 10:3-56.

Franzmeier, D. P., E. J. Pederson, T. J. Longwell, J. C. Byrne, and C. K. Losche (1969). Properties of some soils in the Cumberland Plateau as related to slope aspect and position. *Soil Sci. Soc. Am. Proc.* 33:755-761.

Frazee, C. J., J. B. Fehrenbacher, and W. C. Krumbein (1970). Loess distribution from a source. *Soil Sci. Soc. Am. Proc.* 34:296-301.

Friedkin, J. F. (1945). *A Laboratory Study of the Meandering of Alluvial Rivers.* U.S. Waterways Exper. Sta., Mississippi River Comm., Vicksburg, Miss.

Fritts, H. C. (1965). Dendrochronology. In *The Quaternary of the United States,* ed. H. E. Wright and D. G. Frey, pp. 871-879. Princeton University Press, Princeton.

Frye, J. C. (1954). Graded slopes in western Kansas. *Kans. Geol. Surv. Bull.* 109:85-96.

———— (1959). Climate and Lester King's "Uniformitarian nature of hillslopes." *J. Geol.* 67:111-113.

Frye, J. C., and A. R. Leonard (1954). Some problems on alluvial terrace mapping. *Am. J. Sci.* 252:242-251.

Frye, J. C., and A. B. Leonard (1957). Ecological interpretations of Pliocene and Pleistocene stratigraphy in the Great Plains region. *Am. J. Sci.* 255:1-11.

Frye, J. C., and H. B. Willman (1958). Permafrost features near the Wisconsin glacial margin in Illinois. *Am. J. Sci.* 256:518-524.

———— (1960). *Classification of the Wisconsinan stage in the Lake Michigan Glacial Lobe.* Ill. Geological Survey circ. 285.

Frye, J. C., P. R. Shaffer, H. B. Willman, and G. E. Ekblaw (1960). Accretion-gley and the gumbotil dilemma. *Am. J. Sci.* 258:185-190.

Frye, J. C., H. B. Willman, M. Rubin, and R. F. Black (1968). *Definition of the Wisconsinan Stage,* pp. 1-22. U.S. Geological Survey bull. 1274-E.

Frye, J. C., H. D. Glass, and H. B. Willman (1968). *Mineral Zonation of Woodfordian Loesses of Illinois.* Ill. Geological Survey circ. 427.

Geyl, W. F. (1961). Morphometric analysis and the worldwide occurrence of stepped erosion surfaces. *J. Geol.* 69:388-416.

Ghose, B., and S. Pandey (1963). Quantitative geomorphology of drainage basins. *J. Indian Soc. Soil Sci.* 11:259-274.

Gile, L. H. (1961). A classification of *ca* horizons in soils of a desert region, Dona Ana County, New Mexico. *Soil Sci. Soc. Am. Proc.* 25:52-61.

———— (1966). Coppice dunes and the Rotura soil. *Soil Sci. Soc. Am. Proc.* 30:657-660.

———— (1970). Soils of the Rio Grande Valley border in southern New Mexico. *Soil Sci. Soc. Am. Proc.* 34:465-472.

Gile, L. H., F. F. Peterson, and R. B. Grossman (1965). The *K* horizon: a master soil horizon of carbonate accumulation. *Soil Sci.* 99:74-82.

Gile, L. H., J. W. Hawley, and R. B. Grossman (1970). *Distribution and Genesis of Soils and Geomorphic Surfaces in a Desert Region of Southern New Mexico.* Soil-Geomorphology Field Conference Guidebook. Soil Science Society of America, Madison, Wisconsin.

Gilluly, J. (1937). Physiography of the Ajo region, Arizona. *Geol. Soc. Am. Bull.* 48:323-348.

Gleason, P. J. (1972). *Holocene Sedimentation in the Everglades and Saline Tidal Lands.* Field Conference Guidebook, Second Biennial Meeting. American Quaternary Association, Miami.

Glen, J. W. (1952). Experiments on the deformation of ice. *J. Glaciol.* 2:111-114.

————(1963). The rheology of ice. In *Ice and Snow: Properties, Processes, and Applications,* ed. W. D. Kingery, pp. 3-7. MIT Press, Cambridge.

Glenn, J. L., A. R. Dahl, C. J. Roy, and D. T. Davidson (1960). *Missouri River Studies: Alluvial Morphology and Engineering Soil Classification.* Iowa Engineering Experiment Station Report, Ames, Iowa.

Glock, W. S. (1931). The development of drainage systems: a synoptic review. *Geogr. Rev.* 21:475-482.

Goldich, S. S. (1938). A study in rock weathering. *J. Geol.* 46:17-58.

Goldthwait, R. P. (1958). Wisconsin age forests in western Ohio: I, age and glacial events. *Ohio J. Sci.* 58:209-219.

Goldthwait, R. P., ed. (1971). *Till: A symposium.* Ohio State University Press, Columbus.

Graham, E. R. (1941). Acid clay: an agent in chemical weathering. *J. Geol.* 49:392-401.

———— (1950). The plagioclase feldspars as an index to soil weathering. *Soil Sci. Soc. Am. Proc.* 14: 300-302.

Gravenor, C. P. (1953). The origin of drumlins. *Am. J. Sci.* 251:674-681.

Gwynne, C. S. (1942). Swell-and-swale pattern of the Wisconsin Drift Plain in Iowa. *J. Geol.* 50: 200-208.

———— (1951). Minor moraines in South Dakota and Minnesota. *Geol. Soc. Am. Bull.* 62:233-250.

Hack, J. T. (1941). Dunes of western Navajo country. *Geogr. Rev.* 31:240-263.

———— (1957). *Studies of Longitudinal Stream Profiles in Virginia and Maryland.* U.S. Geological Survey prof. paper 294-B.

_____ (1960). Interpretation of erosional topography in humid temperate regions, pp. 80-97. *Am. J. Sci.* 258-A.

_____ (1965). *Geomorphology of the Shenandoah Valley, Virginia and West Virginia, and Origin of Residual Ore Deposits.* U.S. Geological Survey prof. paper 484.

Hadley, R. D. (1967). Pediments and pediment-forming processes. *J. Geol. Educ.* 15:83-89.

Haines, W. B. (1923). The volume changes associated with variations of water content in soil. *J. Agric. Sci.* 13:296-310.

Handy, R. L. (1972). Alluvial cutoff dating from subsequent growth of a meander. *Geol. Soc. Am. Bull.* 83:475-480.

Hanna, R. M., and O. W. Bidwell (1955). The relation of certain loessial soils of northeastern Kansas to the texture of the underlying loess. *Soil Sci. Soc. Am. Proc.* 19:354-359.

Harper, W. G., and L. H. Smith (1932). *Soil Survey of the Lovington Area, New Mexico,* ser. 1932, no. 2. U.S. Department of Agriculture, Bureau of Chemistry and Soils.

Harward, M. E., and C. T. Youngberg (1969). Soils from Mazama ash in Oregon: identification, distribution, and properties. In *Pedology and Quaternary Research,* ed. S. Pawluk, pp. 163-178. University of Alberta Press, Edmonton.

Hawley, J. W., and F. E. Kottlowski (1969). Quaternary geology of the south-central New Mexico border region. In *Border Stratigraphy Symposium,* ed. F. E. Kottlowski and D. V. Lemone, pp. 89-115. N.M. Bureau of Mines and Mineral Resources circ. 104.

Hay, R. L., and A. Ijima (1968). Nature and origin of palagonite tuffs of the Honolulu group on Oahu, Hawaii. In *Studies in Volcanology,* ed. R. R. Coats, R. L. Hay, and C. A. Anderson. *Geological Society of America Memoir* 116:331-376.

Hays, O. E., and O. J. Attoe (1957). *Control of runoff and erosion on Almena silt loam in Wisconsin.* U.S. Department of Agriculture, Res. Serv., ARS 41-16.

Hjulström, F. (1935). Studies of the morphological activity of rivers as illustrated by the River Fyris. *Univ. Uppsala Geol. Inst. Bull.* 25:221-557.

Holmes, C. D. (1955). Geomorphic development in humid and arid regions: a synthesis. *Am. J. Sci.* 253:377-390.

Hooke, R. L. (1967). Processes on arid-region alluvial fans. *J. Geol.* 75:438-460.

Horner, S. E. (1953). *Landslides and Engineering Practice.* Highway Research Board Committee on Landslide Investigations, National Research Council, Washington, D.C.

Horton, R. E. (1945). Erosional development of streams and their drainage basins: hydrophysical approach to quantitative morphology. *Geol. Soc. Am. Bull.* 56:275-370.

Howard, A. D. (1942). Pediment passes and the pediment problem. *J. Geomorphol.* 5:3-31, 95-136.

_____ (1959). Numerical systems of terrace nomenclature: a critique. *J. Geol.* 67:239-243.

_____ (1967). Drainage analysis in geologic interpretation: a summation. *Am. Assoc. Petrol. Geol. Bull.* 51:2246-2259.

Hugie, V. K., and H. B. Passey (1964). Soil surface patterns of semiarid soils in northern Utah, southern Idaho, and northeastern Nevada. *Soil Sci. Soc. Am. Proc.* 28:786-792.

Hutton, C. E. (1947). Studies of loess-derived soils in southwestern Iowa. *Soil Sci. Soc. Am. Proc.* 12:424-431.

Inman, D. L., and J. Filloux (1960). Beach cycles related to tide and local wind wave regime. *J. Geol.* 68:225-231.

Ireland, H. A., C. F. S. Sharpe, and D. H. Eargle (1939). *Principles of Gully Erosion in the Piedmont of South Carolina.* U.S. Department of Agriculture tech bull. 633.

Jachowski, R. A., ed. (1966). *Shore Protection, Planning, and Design.* 3d ed. U.S. Army Corps of Engineers, Coastal Engineering Research Center tech. rept. 4.

Jackson, M. L. (1958). *Soil Chemical Analysis.* Prentice-Hall, Englewood Cliffs, N.J.

_____ (1964). Chemical composition of soils. In *Chemistry of the Soil,* ed. F. E. Bear, pp. 71-141. Van Nostrand Reinhold Co., New York.

Jackson, M. L., and G. D. Sherman (1953). *Chemical Weathering of Minerals in Soils,* pp. 219-318. Vol. 5. Advances in Agronomy. Academic Press, New York.

Jenny, H. (1941). *Factors of Soil Formation.* McGraw-Hill, New York.

Johnson, D. W. (1919). *Shore Processes and Shoreline Development.* John Wiley & Sons, New York.

Juang, T. C., and G. Uehara (1968). Mica genesis in Hawaiian soils. *Soil Sci. Soc. Am. Proc.* 32:31-35.

Junge, C. E., and R. T. Werby (1958). The concentration of chloride, sodium, potassium, calcium, and sulfate in rain water over the United States. *J. Meteorol.* 15:417-425.

Kay, G. F. (1931). Classification and duration of the

Pleistocene period. *Geol. Soc. Am. Bull.* 42: 425-466.

Kay, G. F., and J. R. Graham (1943). The Illinoian and post-Illinoian Pleistocene geology of Iowa. *Iowa Geol. Surv. Annu. Rept.* 38:1-262.

Kellogg, C. E. (1941). Climate and soil. In *Climate and man*, *U.S. Department of Agriculture Yearbook of Agriculture*, pp. 265-291.

Kempton, J. P., and D. L. Gross (1971). Rate of advance of the Woodfordian (Late Wisconsinan) glacial margin in Illinois: stratigraphic and radiocarbon evidence. *Geol. Soc. Am. Bull.* 82:3245-3250.

King, C. A. M. (1972). *Beaches and Coasts.* 2d ed. Edward Arnold, London.

King, L. C. (1951). *South African Scenery.* 2d ed. Oliver & Boyd, Edinburgh.

_____(1957). The uniformitarian nature of hillslopes. *Trans. Edinburgh Geol. Soc.* 17:81-102.

King, P. B. (1965). Tectonics of Quaternary time in middle North America. In *The Quaternary of the United States,* ed. H. E. Wright and D. G. Frey, pp. 831-870. Princeton University Press, Princeton.

Kirkby, M. J. (1967). Measurement and theory of soil creep. *J. Geol.* 75:359-378.

Kleiss, H. J. (1970). Loess distribution along the Illinois soil-development sequence. *Soil Sci.* 115:194-198.

_____(1973). Hillslope sedimentation and soil formation in northeastern Iowa. *Soil Sci. Soc. Am. Proc.* 34:281-290.

Kulp, J. L. (1961). Geologic time scale. *Science* 133:1105-1114.

Kunkle, G. R. (1968). *A Hydrogeologic Study of the Ground-Water Reservoirs Contributing Base Run-off to Four Mile Creek, East-Central Iowa.* U.S. Geological Survey Water Supply paper 1839-O.

Kunze, G. W., and E. H. Templin (1956). Houston Black Clay, the type Grumosol: II. Mineralogical and chemical characterization. *Soil Sci. Soc. Am. Proc.* 20:91-96.

Lachenbruch, A. H. (1962). *Mechanics of Thermal Contraction Cracks and Ice-Wedge Polygons in Permafrost.* Geological Society of America spec. paper 70.

Langbein, W. B. and others, (1947). Topographic characteristics of drainage basins, pp. 125-155. U.S. Geological Survey Water Supply paper 968-C.

Langbein, W. B., and L. B. Leopold (1964). Quasi-equilibrium states in channel morphology. *Am. J. Sci.* 262:782-794.

Lawrence, D. B., and J. A. Elson (1953). Periodicity of deglaciation in North America since the Late Wis-

consin maximum. *Geogr. Ann.* 35:83-104.

Lehmann, O. (1933). Morphologische Theorie der Verwitterung von Steinschlagwänden. *Vierteljahrsheft Naturforochende Gesellschaft,* Zurich, Switzerland. 78:83-126.

Leighton, M. M. (1933). The naming of the subdivisions of the Wisconsin glacial age. *Science,* n.s. 77:168.

_____(1957). The Cary-Mankato-Valders problem. *J. Geol.* 65:108-111.

Leighton, M. M., and P. MacClintock (1930). Weathered zones of drift sheets of Illinois. *J. Geol.* 38:28-53.

Leighton, M. M., and H. B. Willman (1950). Loess formations of the Mississippi Valley. *J. Geol.* 58:599-623.

Leopold, L. B. (1951). Pleistocene climates in New Mexico. *Am. J. Sci.* 249:152-168.

Leopold, L. B., and T. Maddock (1953). *The Hydraulic Geometry of Stream Channels and Some Physiographic Implications.* U.S. Geological Survey prof. paper 252.

Leopold, L. B., and J. P. Miller (1954). *A Postglacial Chronology for Some Alluvial Valleys in Wyoming.* U.S. Geological Survey Water Supply paper 1261.

Leopold, L. B., and M. G. Wolman (1957). *River Channel Patterns: Braided, Meandering, Straight,* pp. 39-85. U.S. Geological Survey prof. paper 282-B.

_____(1960). River meanders. *Geol. Soc. Am. Bull.* 59:769-794.

Leopold, L. B., and W. B. Langbein (1962). *The Concept of Entropy in Landscape Evolution.* U.S. Geological Survey prof. paper 500-A.

Leopold, L. B., M. G. Wolman, and J. P. Miller (1964). *Fluvial Processes in Geomorphology.* W. H. Freeman, San Francisco.

Lepersonne, J. (1949). Le fossé tectonique Lac Albert-Semliki-Lac Edouard. *Société Géologique de Belgique Mémoires* Tome 72, fascicule 1.

_____(1956). Les aplanissements d'erosion dur nord-est du Congo belge et des régions voisines. *Académie Royale des Sciences Coloniales Mémoires* Tome 4, fascicule 7.

Lévêque, A. (1969). Le problème des sols à nappes de gravats au Togo. *Cah. O.R.S.T.O.M.* (Paris), sér. pédol. 7:43-69.

Libby, W. F. (1965). *Radiocarbon Dating.* 2d ed. University of Chicago Press, Chicago.

Lohnes, R. A., and R. L. Handy (1968). Slope angles in friable loess. *J. Geol.* 76:247-258.

Long, J. T., and R. P. Sharp (1964). Barchan-dune movement in Imperial Valley, California. *Geol. Soc. Am. Bull.* 75:149-156.

Looman, H. (1956). Observations about some differential equations concerning recession of mountain slopes I & II. *Proc. Kon. Ned. Akad. Wetensch.* 59:259-271, 272-284.

Lustig, L. K. (1965). *Clastic Sedimentation in Deep Springs Valley, California,* pp. 131-192. U.S. Geological Survey prof. paper 352-F.

Mabbutt, J. A. (1966). Mantle-controlled planation of pediments. *Am. J. Sci.* 264:78-91.

McCoy, R. M. (1971). Rapid measurement of drainage density. *Geol. Soc. Am. Bull.* 82:757-762.

McFarlan, E. (1961). Radiocarbon dating of late Quaternary deposits, south Louisiana. *Geol. Soc. Am. Bull.* 72:129-158.

McGee, W J (1897). Sheetflood erosion. *Geol. Soc. Am. Bull.* 8:87-112.

Mackin, J. H. (1948). Concept of the graded river. *Geol. Soc. Am. Bull.* 59:464-511.

Mammerickx, J. (1964). Quantitative observations on pediments in the Mojave and Sonoran Deserts (southwestern United States). *Am. J. Sci.* 262: 417-435.

Marbut, C. F. (1922). Soil classification. In *Life and Work of C. F. Marbut,* ed. H. H. Krusekopf, pp. 85-94. Artcraft Press, Columbia, Mo.

_____ (1935). Soils of the United States. In *Atlas of American Agriculture,* pt. 3. U.S. Department of Agriculture, Bureau of Chemistry and Soils.

_____ (1951). *Soils: Their Genesis and Classification.* Soil Science Society of America, Madison, Wis.

Marshall, C. E. (1964). *The Physical Chemistry and Mineralogy of Soils.* John Wiley & Sons, New York.

Matthai, H. F. (1967). Measurement of peak discharge at width contractions by indirect methods. *U.S. Geol. Surv. Tech. Water Res. Invest.,* book 3, chap. A4.

Meier, M. F. (1960). *Mode of Flow of Saskatchewan Glacier, Alberta, Canada.* U.S. Geological Survey prof. paper 351.

_____ (1965). Glaciers and climate. In *The Quaternary of the United States,* ed. H. E. Wright and D. G. Frey, pp. 795-805. Princeton University Press, Princeton.

Meldrum, H. R., D. E. Perfect, and C. A. Mogen (1941). *Soil Survey, Story County, Iowa,* ser. 1936, no. 9. U.S. Department of Agriculture, Bureau of Plant Ind.

Mellor, M. (1967). *Mass Economy of Antarctica: Measurements at Mawson, 1957.* Australian National Antarctic Research Expedition Scientific Report, Melbourne, publ. 97.

Melton, F. A. (1940). A tentative classification of sand dunes: its application to dune history in the southern High Plains. *J. Geol.* 48:113-174.

Melton, M. A. (1957). *An Analysis of the Relations among Elements of Climate, Surface Properties, and Geomorphology.* Columbia University, Dept. of Geology, ONR tech. rept. 11, proj. NR 389-042.

_____ (1965). The geomorphic and paleoclimatic significance of alluvial deposits in southern Arizona. *J. Geol.* 73:1-38.

Merritt, R. S., and E. H. Muller (1959). Depth of leaching in relation to carbonate content of till in central New York state. *Am. J. Sci.* 257:465-480.

Miller, G. A. (1971). *A Geomorphic, Hydrologic, and Pedologic Study of the Iowa Great Lakes Area.* Iowa Conservation Commission, Des Moines, open file report.

Milne, G. (1936a). A provisional soil map of East Africa. East African Agriculture Research Station, Amani Memoirs, Tanganyika Territory.

_____ (1936b). Normal erosion as a factor in soil profile development. *Nature* 138:548.

Moore, D. R. (1972). *Coral Reefs, New and Old.* Field Conference Guidebook, Second Biennial Meeting. American Quaternary Association, Miami.

Morisawa, M. E. (1959). *Relation of Quantitative Geomorphology to Stream Flow in Representative Watersheds of the Appalachian Plateau Province.* Columbia University, Department of Geology, ONR tech. rept. 20., proj. NR 389-042.

_____ (1968). *Streams: Their Dynamics and Morphology.* McGraw-Hill, New York.

Morrison, R. B. (1965). Quaternary geology of the Great Basin. In *The Quaternary of the United States,* ed. H. E. Wright and D. G. Frey, pp. 265-285. Princeton University Press, Princeton.

Mortensen, J. L., and F. L. Himes (1964). Soil organic matter. In *Chemistry of the Soil,* ed. F. E. Bear, pp. 206-241. Van Nostrand Reinhold Co., New York.

Muller, E. H. (1965). Quaternary geology of New York. In *The Quaternary of the United States,* ed. H. E. Wright and D. G. Frey, pp. 99-112. Princeton University Press, Princeton.

Musgrave, G. W. (1947). The quantitative evaluation of factors in water erosion: a first approximation. *J. Soil and Water Conserv.* 2:133-138.

Myers, R. E. (1963). *Floods at Des Moines, Iowa.* U.S. Geological Survey Hydrologic Atlas HA-53.

Norris, R. M. (1966). Barchan dunes of Imperial Valley, California. *J. Geol.* 74:292-306.

North Central Region-3 Technical Committee on Soil Survey (1960). *Soils of the North-Central Region of the United States.* Wis. Agricultural Experiment Station bull. 544.

Nye, J. F. (1952). The mechanics of glacier flow. *J. Glaciol.* 2:82-93.

Odynsky, W. (1958). U-shaped dunes and effective wind directions in Alberta. *Can. J. Soil Sci.* 38: 56-62.

Olson, E. A., and W. S. Broecker (1958). Sample contamination and reliability of radiocarbon dates. *N.Y. Acad. Sci. Trans.* 20:593-604.

Oschwald, W. R., F. F. Riecken, R. I. Dideriksen, W. H. Scholtes, and F. W. Schaller (1965). *Principal Soils of Iowa.* Iowa State University Extension Service spec. rept. 42.

Parizek, E. J., and J. F. Woodruff (1957). Description and origin of stone layers in soils of the southeastern states. *J. Geol.* 65:24-34.

Patton, J. B. (1956). Earthslips in the Allegheny Plateau region. *J. Soil and Water Conserv.* 11, no. 1.

Peltier, L. (1950). The geographic cycle in periglacial regions as it is related to climatic geomorphology. *Assoc. Am. Geogr. Ann.* 40:214-236.

Penck, W. (1924). *Morphological Analysis of Landforms.* Trans. K. C. Boswell and H. Czech, 1953. Macmillan Co., London.

Perkins, R. D., and P. Enos (1968). Hurricane Betsy in the Florida–Bahama area: geologic effects and comparison with Hurricane Donna. *J. Geol.* 76:710-717.

Pettijohn, F. J. (1941). Persistence of heavy minerals and geologic age. *J. Geol.* 49:610-625.

Péwé, T. L. (1965). Notes on the physical environment of Alaska. In *Proceedings 15th Alaska Science Conference,* Alaska Division American Association for Advancement of Science, Fairbanks, pp. 293-310.

――――(1966). *Permafrost and Its Effect on Life in the North.* Oregon State University Press, Corvallis.

―――― (1970). Altiplanation terraces of early Quaternary age near Fairbanks, Alaska. *Acta Geogr. Lodziensia* 24:357-363.

Péwé, T. L., R. E. Church, and M. J. Andresen (1969). *Origin and Paleoclimatic Significance of Large-scale Patterned Ground in the Donelly Dome Area, Alaska.* Geological Society of America spec. paper 103.

Polach, H. A., and J. Golson (1966). *Collection of Specimens for Radiocarbon Dating and Interpretation of Results,* man. 2. Australian Institute for Aboriginal Studies, Canberra.

Polach, H. A., J. Chappell, and J. F. Lovering (1969). ANU radiocarbon date list III. *Radiocarbon* 11:245-262.

Price, W. A. (1958). *Sedimentology and Quaternary Geomorphology of South Texas,* pp. 41-75. Guidebook Spring Field Trip Corpus Christi Geological Society., Corpus Christi.

Prill, R. C., and F. F. Riecken (1958). Variations in forest-derived soils formed from Kansan till in southern and southeastern Iowa. *Soil Sci. Soc. Am. Proc.* 22:70-75.

Quinn, J. H. (1965). Monadnocks, divides, and Ozark physiography. *Ark. Acad. Sci. Proc.* 19:90-97.

Reger, R. D., and T. L. Péwé (1969). Lichenometric dating in the central Alaska Range. In *The Periglacial Environment, Past and Present,* ed. T. L. Péwé, pp. 223-247. McGill-Queen's University Press, Montreal.

Rex, R. W., J. K. Syers, M. L. Jackson, and R. N. Clayton (1969). Eolian origin of quartz in soils of Hawaiian Islands and in Pacific pelagic sediments. *Science* 163:277-279.

Rich, J. L. (1935). Origin and evolution of rock fans and pediments. *Geol. Soc. Am. Bull.* 46:999-1024.

Richards, L. A. (1954). *Diagnosis and Improvement of Saline and Alkali Soils,* U.S. Department of Agriculture handbook 60.

Richmond, G. M. (1965). Glaciation of the Rocky Mountains. In *The Quaternary of the United States,* ed. H. E. Wright and D. G. Frey, pp. 217-230. Princeton University Press, Princeton.

Richmond, G. M., and J. C. Frye (1957). Note 19: status of soils in stratigraphic nomenclature. *Am. Assoc. Petrol. Geol. Bull.* 41:758-763.

Riecken, F. F., and G. D. Smith (1949). Lower categories of soil classification: family, series, type, and phase. *Soil Sci.* 67:107-115.

Riquier, J. (1969). Contribution a l'étude des "Stone-lines" en régions tropicale et equatoriale. *Cah. O.R.S.T.O.M.* (Paris), sér. pédol. 7:71-110.

Robinson, G. W. (1949). *Soils: Their Origin, Constitution, and Classification.* Thomas Murby & Co., London.

Rohdenburg, H. (1969). Slope pedimentation and climatic change as principal factors of planation and scarp development in tropical Africa. *Univ. Göttinger Bodenk. Ber.* 10:127-133.

Rosholt, J. N., C. Emiliani, J. Geiss, F. F. Koczy, and P. J. Wangersky (1961). Absolute dating of deep-sea cores by the Pa^{231}/Th^{230} method. *J. Geol.* 69: 162-185.

Ruhe, R. V. (1950). Graphic analysis of drift topographies. *Am. J. Sci.* 248:435-443.

――――(1952). Topographic Discontinuities of the Des Moines Lobe. *Am. J. Sci.* 250:46-56.

―――― (1954). Erosion surfaces of central African interior

high plateaus. *Publ. de l'I.N.E.A.C.* (Brussels), sér. sci. 59.

_____(1956a). Landscape evolution in the High Ituri, Belgian Congo. *Publ. de l'I.N.E.A.C.* (Brussels), sér. sci. 66.

_____ (1956b). Geomorphic surfaces and the nature of soils. *Soil Sci.* 82:441-455.

_____ (1959). Stone lines in soils. *Soil Sci.* 87:223-231.

_____ (1960). Elements of the soil landscape. Transactions 7th International Congress of Soil Science, vol. 4, pp. 165-170. Madison, Wis.

_____ (1964a). Landscape morphology and alluvial deposits in southern New Mexico. *Ann. Assoc. Am. Geogr.* 54:147-159.

_____(1964b). An estimate of paleoclimate in Oahu, Hawaii. *Am. J. Sci.* 262:1098-1115.

_____(1965a). Quaternary paleopedology. In *The Quaternary of the United States*, ed. H. E. Wright and D. G. Frey, pp. 755-764. Princeton University Press, Princeton.

_____(1965b). Relation of fluctuations of sea level to soil genesis in the Quaternary. *Soil Sci.* 99:23-29.

_____(1967). Geomorphic surfaces and surficial deposits in southern New Mexico. *N.M. Bureau of Mines and Mineral Resources Memoir* no. 18.

_____ (1968). Identification of paleosols in loess deposits in the United States. In *Loess and Related Eolian Deposits of the World*, ed. C. B. Schultz and J. C. Frye, pp. 49-65. University of Nebraska Press, Lincoln.

_____(1969a). *Quaternary Landscapes in Iowa.* Iowa State University Press, Ames.

_____(1969b). Application of pedology to Quaternary research. In *Pedology and Quaternary Research*, ed. S. Pawluk, pp. 1-23. University of Alberta Press, Edmonton.

_____(1970). Soils, paleosols, and environment. In *Pleistocene and Recent Environments of the Central Great Plains*, ed. W. Dort and J. K. Jones, pp. 37-52. University of Kansas Press, Lawrence.

_____ (1971). Stream regimen and man's manipulation. In *Environmental Geomorphology*, ed. D. R. Coates, pp. 9-23. State University of New York, Binghamton, Publications in Geomorphology.

_____(1973). Background of model for loess-derived soils in the upper Mississippi River basin. *Soil Sci.* 115:250-253.

Ruhe, R. V., and R. B. Daniels (1958). Soils, paleosols, and soil-horizon nomenclature. *Soil Sci. Soc. Am. Proc.* 22:66-69.

_____ (1965). Landscape erosion: geologic and historic. *J. Soil and Water Conserv.* 20:52-57.

Ruhe, R. V., and J. G. Cady (1967). *The Relation of Pleistocene Geology and Soils between Bentley and Adair in Southwestern Iowa*, pp. 1-92. U.S. Department of Agriculture tech. bull. 1349.

Ruhe, R. V., J. G. Cady, and R. S. Gomez (1961). Paleosols of Bermuda. *Geol. Soc. Am. Bull.* 72:1121-1142.

Ruhe, R. V., J. M. Williams, and E. L. Hill (1965a). Shorelines and submarine shelves, Oahu, Hawaii. *J. Geol.* 73:485-497.

Ruhe, R. V., J. M. Williams, R. C. Shuman, and E. L. Hill (1965b). Nature of soil parent materials in Ewa-Waipahu area, Oahu, Hawaii. *Soil Sci. Soc. Am. Proc.* 29:282-287.

Ruhe, R. V., and P. H. Walker (1968). Hillslope models and soil formation: I, Open systems. Transactions 9th International Congress of Soil Science, vol. 4, pp. 551-560. Adelaide, Australia.

Ruhe, R. V., W. P. Dietz, T. E. Fenton, and G. F. Hall (1968). *Iowan Drift Problem, Northeastern Iowa.* Iowa Geological Survey rept. invest. 7.

Ruhe, R. V., G. A. Miller, and W. J. Vreeken (1971). Paleosols, loess sedimentation, and soil stratigraphy. In *Paleopedology: Origin, Nature, and Dating of Paleosols*, ed. D. H. Yaalon, pp. 41-60. Hebrew Universities Press, Jerusalem.

Russell, R. J. (1940). Lower Mississippi Valley loess. *Geol. Soc. Am. Bull.* 55:1-40.

Ruxton, B. P., and L. Berry (1957). Weathering of granite and associated erosional features in Hong Kong. *Geol. Soc. Am. Bull.* 68:1263-1292.

Rydlerskaya, M. D., and V. V. Tiscenko (1944). On the cation exchange of humic acid of different soil types. *Pedology* 10:491.

Savigear, R. A. G. (1956). Technique and terminology in investigation of slope forms. *Int. Geogr. Union 1st Rept. Comm. Study Slopes* (Rio de Janeiro), pp. 66-75.

Saxton, K. E., Spomer, R. G., and L. A. Kramer (1971). Hydrology and erosion of loessial watershed. *J. Hydraul. Div., Am. Soc. Civ. Engr. Proc.* 97:1835-1851.

Sayles, R. W. (1931). Bermuda during the Ice Age. *Am. Acad. Arts and Sci. Proc.* 66:381-467.

Scharpenseel, H. W., F. Pietig, and M. A. Tamers (1968). Bonn radiocarbon measurements I. *Radiocarbon* 10:8-28.

Scheidegger, A. E. (1961). *Theoretical Geomorphology.* Springer-Verlag, Berlin.

_____(1970). *Theoretical Geomorphology.* 2d ed. Springer-Verlag, Berlin.

Scheidegger, A. E., and W. B. Langbein (1966). *Probability Concepts in Geomorphology.* U.S. Geological Survey prof. paper 500-C.

Scheidegger, A. E., and P. E. Potter (1968). Textural

studies of grading: volcanic ash falls. *Sedimentology* 11:163-170.

Schumm, S. A. (1954). *The Relation of Drainage Basin Relief to Sediment Loss* 1:216-219. International Association of Hydrology publ. 36.

_____ (1956a). Evolution of drainage systems and slopes in badlands at Perth Amboy, New Jersey. *Geol. Soc. Am. Bull.* 67:597-646.

_____(1956b). The movement of rocks by wind. *J. Sedimentary Petrology* 26:284-286.

_____ (1956c). The role of creep and rainwash on the retreat of badland slopes. *Am. J. Sci.* 254:693-706.

_____(1962). Erosion on miniature pediments in Badlands National Monument, South Dakota. *Geol. Soc. Am. Bull.* 73:719-724.

_____(1963). Sinuosity of alluvial rivers on the Great Plains. *Geol. Soc. Am. Bull.* 74:1089-1100.

_____(1964). Seasonal variations of erosion rates and processes on hillslopes in western Colorado. *Z. Geomorphol.* 5:215-238.

_____(1967a). Meander wavelength of alluvial rivers. *Science* 157:1549-1550.

_____ (1967b). Rates of surficial rock creep on hillslopes in western Colorado. *Science* 155:560-561.

Schwob, H. H. (1953). Iowa floods: magnitude and frequency. *Iowa Highway Research Board Bulletin 1*, Ames.

Segalen, P. (1969). Le remaniement des sols et la mise en place de la "Stone-line" en Afrique. *Cah. O.R.S.T.O.M.* (Paris), sér. pédol. 7:113-127.

Selby, M. J. (1966). Methods of measuring soil creep. *N.Z. J. Hydrol.* 5:54-63.

Sharp, R. P. (1954). Glacier flow: a review. *Geol. Soc. Am. Bull.* 65:821-838.

_____ (1963). Wind ripples. *J. Geol.* 71:617-636.

_____ (1969). Semiquantitative differentiation of glacial moraines near Convict Lake, Sierra Nevada, California. *J. Geol.* 77:68-91.

Sharp, R. P., and L. H. Nobles (1953). Mudflow of 1941 at Wrightwood, southern California. *Geol. Soc. Am. Bull.* 64:547-560.

Sharpe, C. F. S. (1938). *Landslides and Related Phenomena.* Columbia University Press, New York.

Shepard, F. P. (1950). *Longshore-Bars and Longshore-Troughs.* U.S. Army, Corps of Engineers, Beach Erosion Board tech. memo. 15.

Shrader, W. D., M. E. Springer, and N. S. Hall (1953). Soil survey of Holt County, Missouri, ser. 1939, no. 21. U.S. Department of Agriculture, Soil Conservation Service.

Shreve, R. L. (1968). *The Blackhawk Landslide.* Geological Society of America spec. paper 108.

Shumskii, P. A. (1964). *Principles of Structural Glaciology.* Trans. David Kraus. Dover, New York.

Simons, D. B., and E. V. Richardson (1966). *Resistance to Flow in Alluvial Channels.* U.S. Geological Survey prof. paper 422-J.

Simonson, R. W. (1941). Studies of buried soils formed from till in Iowa. *Soil Sci. Soc. Am. Proc.* 6: 373-381.

_____ (1954). Identification and interpretation of buried soils. *Am. J. Sci.* 252:705-732.

_____(1959). Outline of a generalized theory of soil genesis. *Soil Sci. Soc. Am. Proc.* 23:152-170.

Small, R. J. (1972). *The study of Landforms: a Textbook of Geomorphology.* Cambridge University Press, London.

Smalley, I. J., and C. Vita-Finzi (1969). The concept of "system" in the earth sciences, particularly geomorphology. *Geol. Soc. Am. Bull.* 80: 1591-1594.

Smith, G. D. (1942). Illinois loess: variations in its properties and distribution. *Ill. Agric. Exper. Sta. Bull.* 490:139-184.

Smith, H. T. U. (1941). Review of *Dunes of Western Navajo Country*, by John T. Hack. *J. Geomorphol.* 4:250-252.

_____(1949). Physical effects of Pleistocene climatic changes in nonglaciated areas: eolian phenomena, frost action, and stream terracing. *Geol. Soc. Am. Bull.* 60:1485-1516.

_____ (1965). Dune morphology and chronology in central and western Nebraska. *J. Geol.* 73:557-578.

Smith, R. M. (1959). *Some Structural Relationships of Texas Blackland Soils with Special Attention to Shrinkage and Swelling.* U.S. Department of Agriculture, Agricultural Research Service, ARS 41-28.

Soil Survey Laboratory (1967). *Laboratory Methods and Procedures for Collection of Soil Samples.* U.S. Department of Agriculture, Soil Conservation Service, Soil Survey Investigations Report No. 1.

Soil Survey Staff (1951). *Soil Survey Manual.* U.S. Department of Agriculture handbook 18.

_____ (1960). *Soil Classification: A Comprehensive System.* U.S. Department of Agriculture.

Speight, J. G. (1971). Log-normality of slope distributions. *Z. Geomorphol.* 15:290-311.

Stace, H. C. T., G. D. Hubble, R. Brewer, K. H. Northcote, J. R. Sleeman, M. J. Mulcahy, and E. G. Hallsworth (1968). *A Handbook of Australian Soils.* Rellim Technical Publications, Glenside, South Australia.

Stall, J. B., and C. T. Yang (1970). *Hydraulic Geometry of Twelve Selected Stream Systems of the United*

States. University of Illinois Water Research Center Report No. 32.

Stearns, H. T. (1935a). Shore benches on the island of Oahu, Hawaii. *Geol. Soc. Am. Bull.* 46: 1467-1482.

_____(1935b). Pleistocene shorelines on the islands of Oahu and Maui, Hawaii. *Geol. Soc. Am. Bull.* 46:1927-1956.

_____(1961). Eustatic shorelines on Pacific islands. *Z. Geomorphol.* 3:3-16.

Stevenson, I. L. (1964). Biochemistry of soil. In *Chemistry of the Soil*, ed. F. E. Bear, pp. 242-291. Van Nostrand Reinhold Co., New York.

Strahler, A. N. (1950). Equilibrium theory of erosional slopes approached by frequency distribution analysis. *Am. J. Sci.* 248:673-696, 800-814.

_____ (1952). Dynamic basis of geomorphology. *Geol. Soc. Am. Bull.* 63:923-938.

_____(1956). Quantitative slope analysis. *Geol. Soc. Am. Bull.* 67: 571-596.

_____ (1958). Dimensional analysis applied to fluvially eroded landforms. *Geol. Soc. Am. Bull.* 69: 279-300.

_____ (1966). Tidal cycle of changes in an equilibrium beach, Sandy Hook, New Jersey. *J. Geol.* 74: 247-268.

_____(1968). Quantitative geomorphology. In *Encyclopedia of Geomorphology*, ed. R. W. Fairbridge, pp. 898-912. Reinhold Publishing Corp., New York.

Straub, L. G., and L. W. Miller (1935). Silt investigations in the Missouri Basin. In *Missouri River: House Doc. 238*, pp. 1032-1245. 73d Cong., 2d sess.

Stuiver, M., and H. E. Suess (1966). On the relationship between radiocarbon dates and true sample ages. *Radiocarbon* 8:534-540.

Swineford, A. (1949). Source area of Great Plains Pleistocene volcanic ash. *J. Geol.* 57:307-311.

Swineford, A., and J. C. Frye (1946). Petrographic comparison of Pliocene and Pleistocene volcanic ash from western Kansas. *Kans. Geol. Surv. Bull.* 64:1-32.

Taras, M. J., A. E. Greenberg, R. D. Hoak, and M. C. Rand, eds. (1971). *Standard Methods for the Examination of Water and Wastewater.* 13th ed. American Public Health Association, New York.

Tator, B. A. (1953). Pediment characteristics and terminology. *Ann. Assoc. Am. Geogr.* 43:47-53.

Taylor, D. W. (1948). *Fundamentals of Soil Mechanics.* John Wiley & Sons, New York.

Templin, E. H., I. C. Mowery, and G. W. Kunze (1956). Houston Black Clay, the type Grumosol: I, Field morphology and geography. *Soil Sci. Soc. Am. Proc.* 20:88-90.

Thiel, C. A. (1944). The geology and underground waters of southern Minnesota. *Minn. Geol. Surv. Bull.* 31.

Thornbury, W. D. (1940). Weathered zones and glacial chronology in southern Indiana. *J. Geol.* 48: 449-475.

_____ (1954). *Principles of Geomorphology.* John Wiley & Sons, New York.

_____ (1965). *Regional Geomorphology of the United States.* John Wiley & Sons, New York.

_____ (1969). *Principles of Geomorphology.* 2d ed. John Wiley & Sons, New York.

Thornthwaite, C. W. (1931). The climates of North America according to a new classification. *Geogr. Rev.* 21:633-655.

_____(1933). The climates of the earth. *Geogr. Rev.* 23:433-440.

_____ (1941). *Atlas of Climatic Types in the United States*, pp. 1-7. U.S. Department of Agriculture misc. publ. 421.

Thorp, J., and H. T. U. Smith (1952). Pleistocene eolian deposits of the United States, Alaska, and parts of Canada. Geological Society of America map.

Thurber, D. L., W. S. Broecker, R. L. Blanchard, and H. A. Potratz (1965). Uranium-series ages of Pacific atoll coral. *Science* 149:55-58.

Tricart, J., and A. Cailleux (1965). *Introduction à la géomorphologie climatique.* SEDES, Paris.

Troeh, F. R. (1965). Landform equations fitted to contour maps. *Am. J. Sci.* 263:616-627.

Trowbridge, A. C. (1921). *The Erosional History of the Driftless Area*, pp. 7-127. University of Iowa Study in Natural History 9, Iowa City.

Tuan, Y. (1962). Structure, climate, and basin land forms in Arizona and New Mexico. *Ann. Assoc. Am. Geogr.* 52:51-68.

Ulrich, R. (1950). Some physical changes accompanying Prairie, Wiesenboden, and Planosol soil profile development from Peorian loess in southwestern Iowa. *Soil Sci. Soc. Am. Proc.* 14:287-295.

Ulrich, R., R. J. Arkley, R. E. Nelson, and R. J. Wagner (1959). Characteristics and genesis of some soils of San Mateo County, California. *Soil Sci.* 88: 218-227.

U.S. Department of Agriculture, Soil Conservation Service (1969). Distribution of principal kinds of soils: orders, suborders, and great groups. In U.S. Geological Survey, *National Atlas*, sheets 85, 86.

Vanoni, V. A. (1971). Sediment transportation mechanics: Q. Genetic classification of valley sediment deposits. J. Hydraulics Div., Am. Soc. Civ. Engr. 95, HY-1:43-53.

Vreeken, W. J. (1968). Stratigraphy, sedimentology, and moisture contents in a small loess watershed in Tama County, Iowa. *Iowa Acad. Sci. Proc.* 75:225-233.

Wagner, W. P. (1970). Ice movement and shoreline modification, Lake Champlain, Vermont. *Geol. Soc. Am. Bull.* 81:117-126.

Wahrhaftig, C. (1965). Stepped topography of the southern Sierra Nevada, California. *Geol. Soc. Am. Bull.* 76:1165-1190.

Walker, P. H. (1966). *Postglacial Environments in Relation to Landscape and Soils*, pp. 838-875. Iowa Agricultural Experiment Station research bulletin 549.

Walker, P. H., and C. A. Hawkins (1958). A study of river terraces and soil development on the Nepean River, N.S.W. *J. Roy. Soc. New South Wales* 91:67-84.

Wascher, H. L., J. D. Alexander, B. W. Ray, A. H. Beavers, and R. T. Odell (1960). *Characteristics of Soils Associated with Glacial Tills in Northeastern Illinois.* Ill. Agricultural Experiment Station bull. 665.

Washburn, A. L. (1956). Classification and origin of patterned ground. *Geol. Soc. Am. Bull.* 67:823-865.

Wayland, E. J. (1934). Peneplains and some other erosional platforms. *Uganda Geol. Surv. Ann. Rept.*, pp. 77-79.

Wayne, W. J. (1967). Periglacial features and climatic gradient in Illinois, Indiana, and western Ohio, east-central United States. In *Quaternary Paleoecology*, ed. E. J. Cushing and H. E. Wright, pp. 393-414. Yale University Press, New Haven.

Wayne, W. J., and J. H. Zumberge (1965). Pleistocene geology of Indiana and Michigan. In *The Quaternary of the United States*, ed. H. E. Wright and D. G. Frey, pp. 63-84. Princeton University Press, Princeton.

Welander, P. (1961). Numerical prediction of storm surges. In *Advances in Geophysics*, ed. H. E. Landsberg and J. Van Mieghern, 8:315-379. Academic Press, New York.

Wells, P. V. (1970). Postglacial vegetational history of the Great Plains. *Science* 167:1574-1582.

Wentworth, C. K., and H. S. Palmer (1925). Eustatic bench of islands of the North Pacific. *Geol. Soc. Am. Bull.* 36:521-544.

Wentworth, C. K., and J. E. Hoffmeister (1939). Geology of Ulupau Head, Oahu. *Geol. Soc. Am. Bull.* 50:1553-1572.

West, E. M. (1960). Drainage for highways and airports. In *Highway Engineering Handbook*, ed. K. B. Woods, D. S. Berry, and W. H. Goetz, pp. 12-1–12-44. McGraw-Hill, New York.

Westgate, J. A., D. G. W. Smith, and H. Nichols (1969). Late Quaternary pyroclastic layers in the Edmonton area, Alberta. In *Pedology and Quaternary Research*, ed. S. Pawluk, pp. 179-186. University of Alberta Press, Edmonton.

Weyl, P. F. (1958). The solution kinetics of calcite. *J. Geol.* 66:163-176.

Weyl, R. (1952). Studies of heavy minerals in soil profiles. *Z. Pfl. Ernahr. Dung. Bodenk.* 57:135-141.

Whipple, W. (1942). Missouri River slope and sediment. *Am. Soc. Civ. Engrs. Trans.* 107:1178-1200.

White, S. E., and M. D. Malcolm (1972). Geomorphology in North American geology departments, 1971. *J. Geol. Educ.* 20:143-147.

White, W. A. (1972). Deep erosion by continental ice sheets. *Geol. Soc. Am. Bull.* 83:1037-1056.

Wilcox, R. E. (1965). Volcanic-ash chronology. In *The Quaternary of the United States*, ed. H. E. Wright and D. G. Frey, pp. 807-816. Princeton University Press, Princeton.

Willden, R., and D. R. Mabey (1961). Giant dessication fissures on the Black Rock and Smoke Creek Deserts, Nevada. *Science* 133:1359-1360.

Williams, G. E., and H. A. Polach (1971). Radiocarbon dating of arid-zone calcareous paleosols. *Geol. Soc. Am. Bull.* 82:3069-3086.

Williams, G. P., and H. P. Guy (1971). Debris avalanches: a geomorphic hazard. In *Environmental Geomorphology*, ed. D. R. Coates, pp. 25-47. State University of New York, Binghamton, Publications in Geomorphology.

Willman, H. B., H. D. Glass, and J. C. Frye (1966). *Mineralogy of Glacial Tills and Their Weathering Profiles in Illinois: Pt. II, Weathering Profiles.* Ill. Geological Survey circ. 400.

Willman, H. B., and J. C. Frye (1970). *Pleistocene Stratigraphy of Illinois.* Ill. Geological Survey bull. 94.

Wischmeier, W. H., and D. D. Smith (1965). *Predicting Rainfall Erosion Loesses from Cropland East of the Rocky Mountains.* U.S. Department of Agriculture handbook 282.

Wolman, M. G. (1971). Evaluating alternative techniques of floodplain mapping. *Water Resources Res.* 7:1383-1392.

Wolman, M. G., and L. B. Leopold (1957). *River Flood-*

plains: Some Observations on Their Formation, pp. 87-109. U.S. Geological Survey prof. paper 282-C.

Wood, A. (1942). The development of hillside slopes. *Geol. Assoc. Proc.* 53:128-138.

Woodring, W. P., M. N. Bramlette, and W. S. W. Kew (1946). *Geology and Paleontology of Palos Verde Hills, California*. U.S. Geological Survey prof. paper 207.

Woodward, S. M., and C. J. Posey (1941). *Hydraulics of Steady Flow in Open Channels*. John Wiley & Sons, New York.

Wright, H. E. (1970). Vegetational history of the Central Plains. In *Pleistocene and Recent Environments of the Central Great Plains*, ed. W. Dort and J. K. Jones, pp. 157-172. University of Kansas Press, Lawrence.

Yaalon, D. H., ed. (1971). *Paleopedology: Origin, Nature, and Dating of Paleosols*. Israel University Press, Jerusalem.

Yasso, W. E. (1971). Forms and cycles in beach erosion and deposition. In *Environmental Geomorphology*, ed. D. R. Coates, pp. 109-137. State University of New York, Binghamton, Publications in Geomorphology.

Yearbook of Agriculture (1938). *Soils and Men*. U.S. Department of Agriculture.

Yehle, L. A. (1954). Soil tongues and their confusion with certain indicators of periglacial climate. *Am. J. Sci.* 252:532-546.

Young, A. (1960). Soil movement by denudational processes on slopes. *Nature* 188:120-122.

Zenkovich, V. P. (1967). *Processes of Coastal Development*, ed. and trans. J. A. Steers. Oliver & Boyd, Edinburgh.

Zingg, A. W. (1940). Degree and length of landslope as it affects soil loss in runoff. *Agric. Engr.* 21:59-64.

Index

micromorphology, 74-77 (fig. 4.5)
normal, 121, 132
organic matter, 28-29, 74
profile, 37-42 (figs. 2.11, 2.12, 2.13), 202
reaction, 14-15
structure, 13-14 (table 1.3), 221-222
survey, 77
texture, 12-13 (fig. 1.8), 25-26
Solifluction, 112, 117, 121 (fig. 6.17)
Solum, 38
Solution, 21
Specific surface, 23-24 (table 2.1)
Spit, 174-175
Statistical method, 15-16
Steady state, 95-97, 109
Steppe, 207
Stepped surfaces, 127 (fig. 7.2), 132-133 (fig. 7.7), 136-141 (figs. 7.13, 7.14, 7.17), 147 (table 7.2)
Stokes' law, 60, 69, 151
Stone line, 127-129 (fig. 7.3)
Stratigraphy, 6-11
glacial stages, 201-204
morphostratigraphic unit, 198
pleistocene, 16 (table 1.5), 201
soil, 202-203
soil-stratigraphic unit, 203
Stream, 47-68
braided, 67-68 (fig. 3.16)
channel, 57-58 (figs. 3.7, 3.8), 61-68 (figs. 3.12, 3.15)
frequency, 91
graded, 63
meander, 64-66 (fig. 3.14)
order, 90-91 (fig. 5.3)
powder, 58-60 (fig. 3.10)
profile, 61-64 (fig. 3.12)
work of, 56, 61
Stress-strain, 22-23, 110, 187-188 (fig. 10.1)
maximum strength, 110
shearing strength, 22, 110, 117
shearing stress, 110, 117, 188
Stripped plain, 140
Structure, 13-14, 105, 125, 140, 159
structural control, 88
structural dome, 88-89 (fig. 5.1)
structural plain, 140
see also Soil
Subsidence, 112 (table 6.4)
Superposition, law of, 222
Surficial geology, 3
deposits, 3, 153 (fig. 8.2)
sediments, 3
Surge, 170, 190
glacier, 190-191 (fig. 10.5)

storm, 170-173 (fig. 9.6)
Systems analysis, 4 (fig. 1.2), 95-97, 148
open system, 95, 101-102 (fig. 6.2), 109
closed system, 95 (fig. 5.6), 109

Taiga, 206
Tectonic movement, 80, 86, 177, 180-182
Temperature efficiency index, 206-207 (figs. 11.1, 11.2), 213
Tephra, 165
Tephrochronology, 165
Terrace, 75-81 (fig. 4.4)
altiplanation, 117
cut, 78-80
fill, 78-80
marine, 178-181
nomenclature, 80 (fig. 4.8)
strath, 79
Texture, see Soil
Thermocline, 185
Tide, 170-172 (fig. 9.5)
Time, 15-18 (tables 1.4, 1.5), 36, 145. See also Dating
Toposequence, 42
Transport, see Sediment
Tsunami, 167
Tundra, 206

Underfit stream, 182

Valley train, 201
Velocity, 49, 61 (fig. 3.11)
critical, 49, 59
current, 173
drag, 150, 156
orbital, 169
threshold, 150-156
vertical-velocity curve, 51-52 (fig. 3.3), 149, 188-189 (fig. 10.2)
wave, 168
Ventifact, 152
Volcano, 4 (fig. 1.1)
volcanic ash, 164-165

Water, 24-25, 47
balance, 47, 214
capillary, 24
chemically combined, 24
gravitational, 24
hygroscopic, 24
Watershed, 87, 133, 136
Wave, 167-171 (figs. 9.1, 9.3)
action, 167-168
cut bench, 173, 178 (fig. 9.11)